SCAN THE CODE TO ACC[ESS] YOUR FREE DIGITAL COPY OF THE NEUROANATOMY COLORING BOOK

SCAN ME

The Neuroanatomy Coloring Book features:

• **The most effective way to skyrocket your neuroanatomical knowledge, all while having fun!**

• Full coverage of the major systems of the human brain to provide context and reinforce visual recognition

• **25+ unique, easy-to-color pages of different neuroanatomical sections with their terminology**

• Large 8.5 by 11-inch single side paper so you can easily remove your coloring

• **Self-quizzing for each page, with convenient same-page answer keys**

THIS BOOK BELONGS TO

TO

_____

TABLE OF CONTENTS

YOGA POSES FOR INTERMEDIATES

YOGA POSES FOR INTERMEDIATES

1. SIDE PLANK POSE

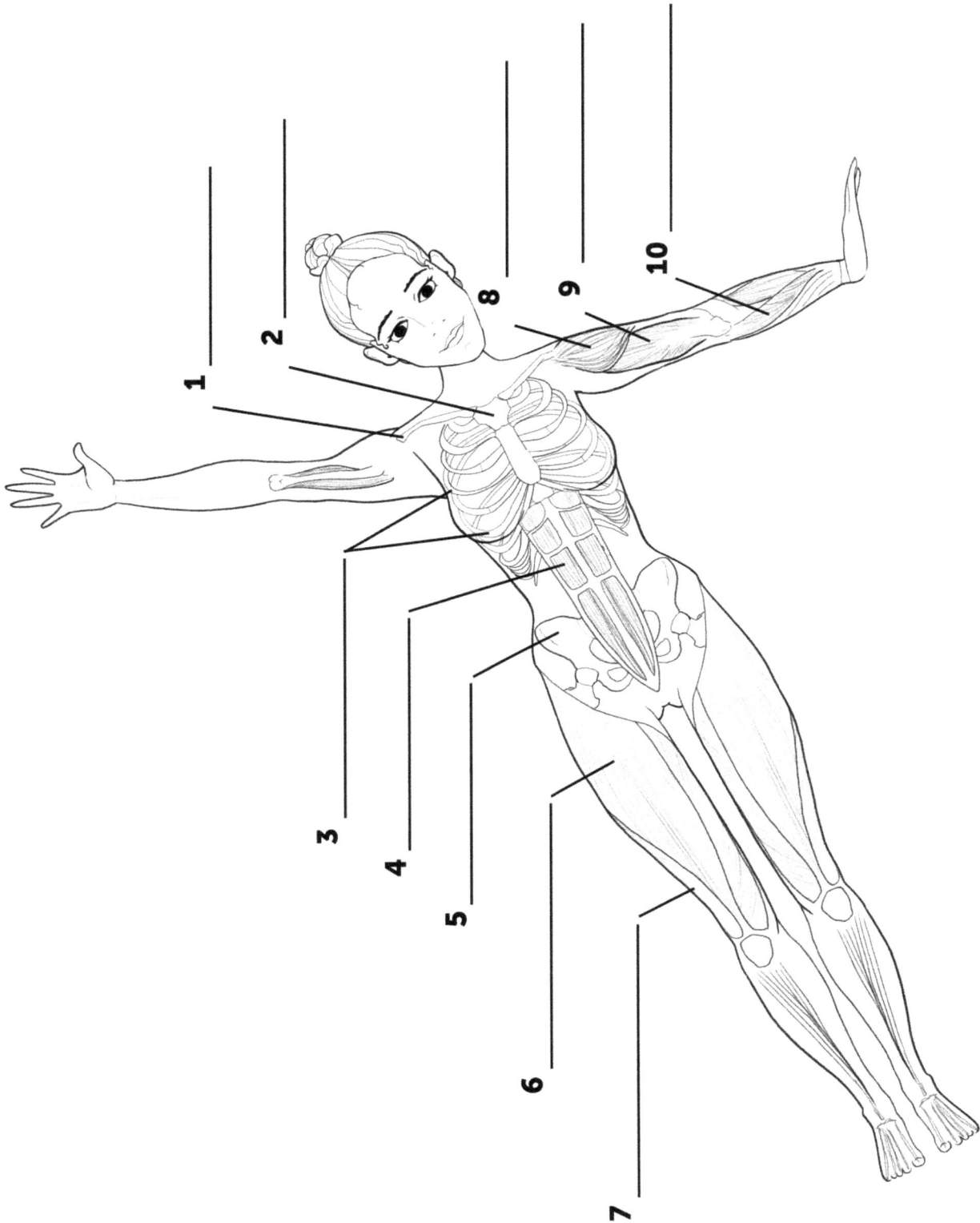

1
2
3
4
5
6
7
8
9
10

1. SIDE PLANK POSE

1. COLLARBONE
2. STERNUM
3. RIBS
4. RECTUS ABDOMINIS
5. PELVIS
6. QUADRICEPS
7. VASTUS LATERALIS
8. DELTOID
9. BICEPS BRACHII
10. PRONATORS

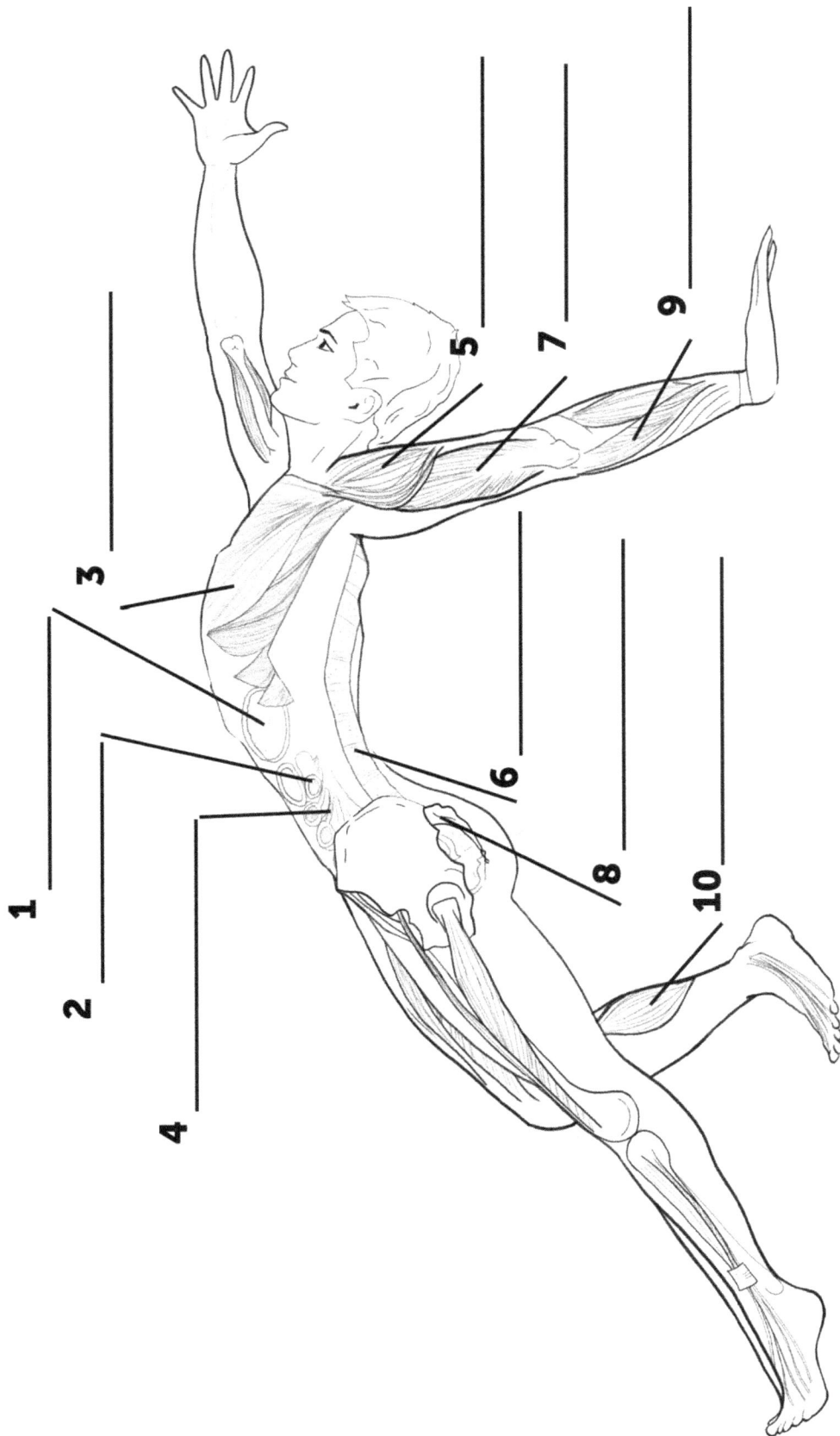

2. WILD THING

1

2

3

4

5

7

9

6

8

10

2. WILD THING

1. STOMACH

2. COILS OF SMALL INTESTINE

3. PECTORALIS MAJOR

4. MESENTERY OF SMALL INTESTINE

5. DELTOID

6. SPINE

7. BICEPS BRACHII

8. SACRUM

9. PRONATORS

10. GASTROCNEMIUS

3. HALF FROG POSE

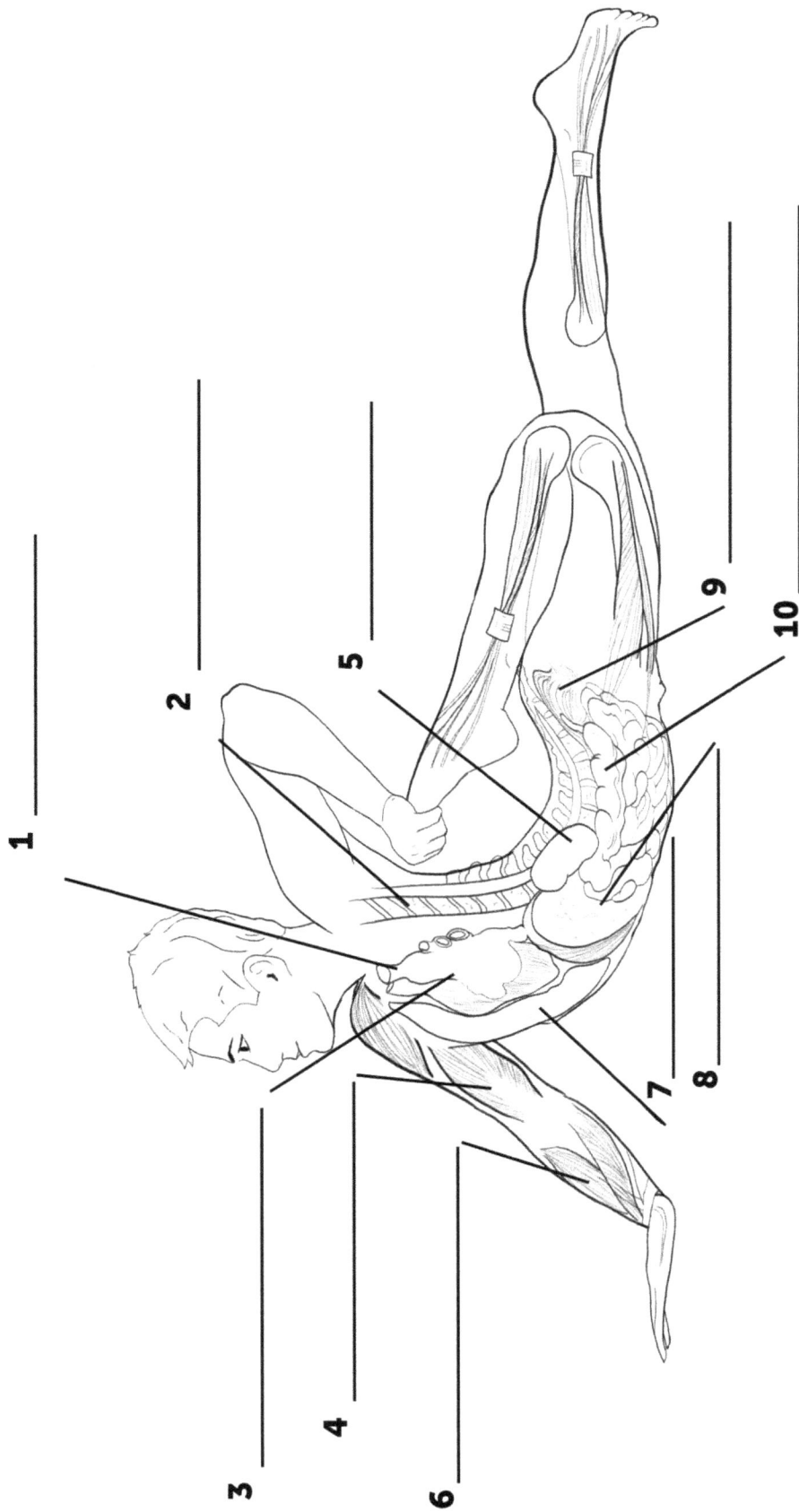

1

2

3

4

5

6

7

8

9

10

3. HALF FROG POSE

1. AORTA

2. SPINE

3. HEART

4. BICEPS BRACHII

5. KIDNEY

6. PRONATORS

7. LUNGS

8. LIVER

9. RECTUM

10. ASCENDING COLON

4. COMPASS POSE

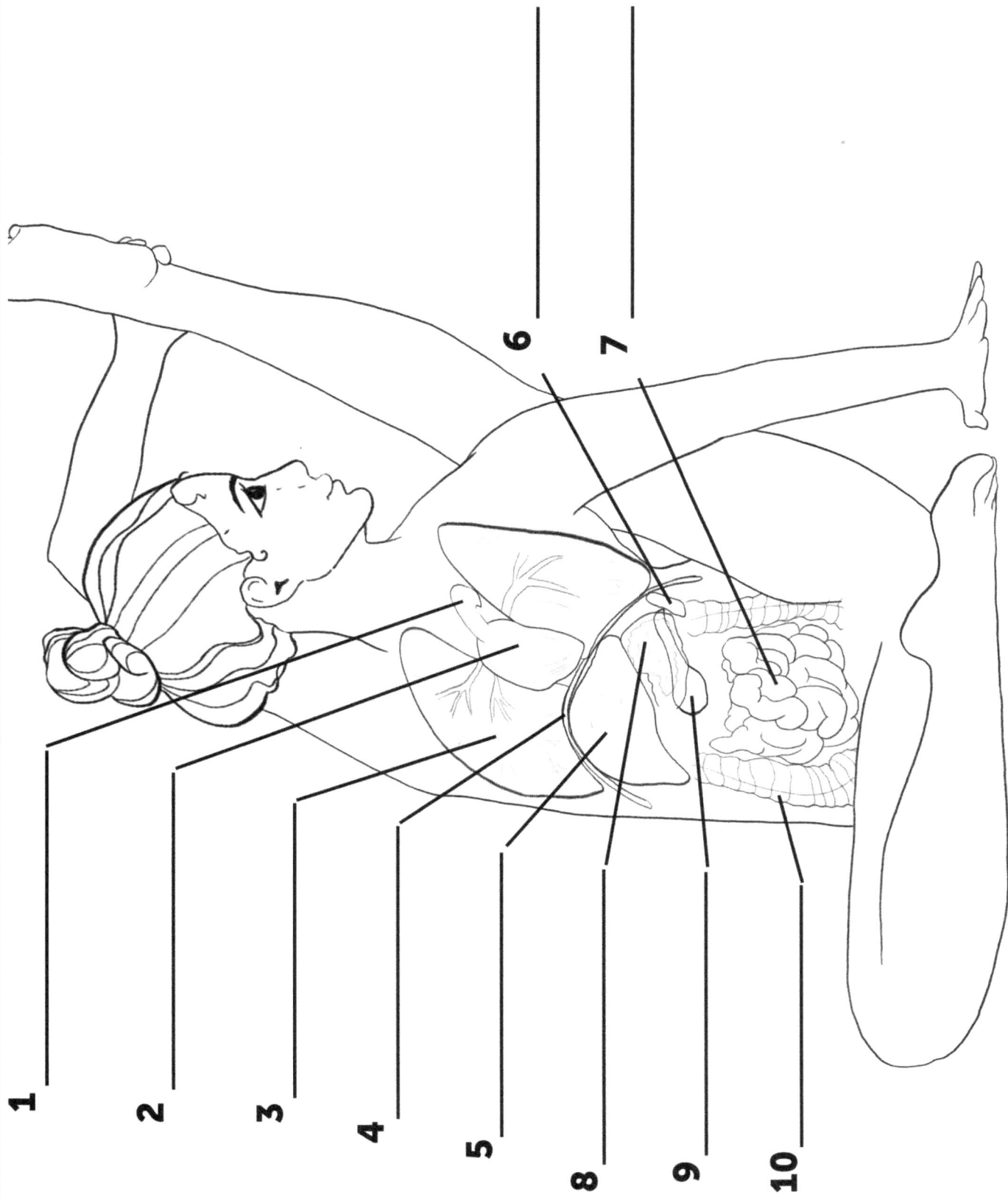

1

2

3

4

5

6

7

8

9

10

4. COMPASS POSE

1. AORTA

2. HEART

3. LUNGS

4. DIAPHRAGM

5. LIVER

6. GALLBLADDER

7. COILS OF SMALL INTESTINE

8. STOMACH

9. PANCREAS

10. ASCENDING COLON

5. MARICHI'S POSE I

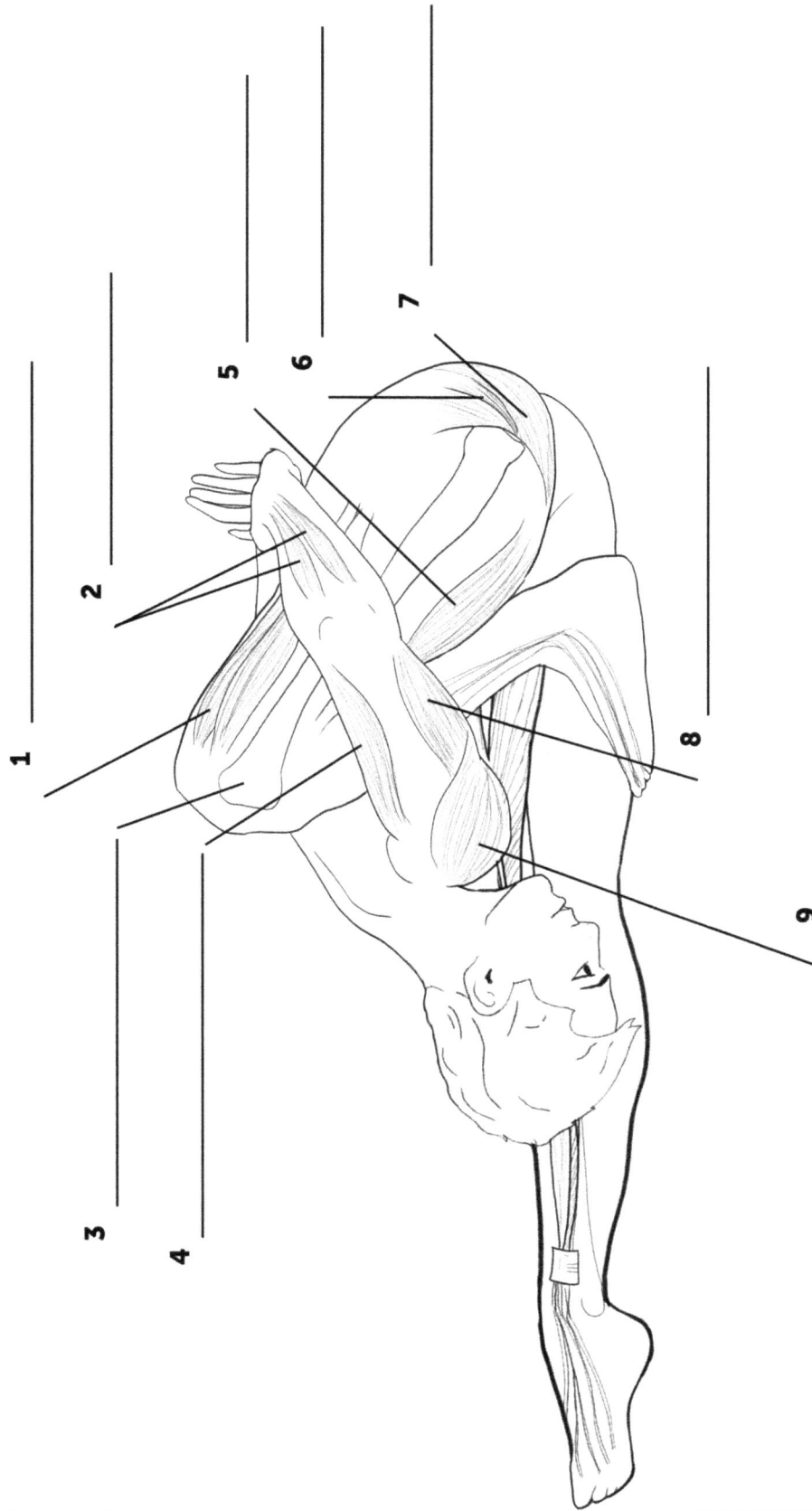

1

2

3

4

5

6

7

8

9

5. MARICHI'S POSE I

1. QUADRICEPS

2. PRONATORS

3. FEMUR

4. BICEPS BRACHII

5. HAMSTRINGS

6. PIRIFORMIS

7. GLUTEUS MAXIMUS

8. TRICEPS BRACHII

9. DELTOID

6. MARICHI'S POSE II

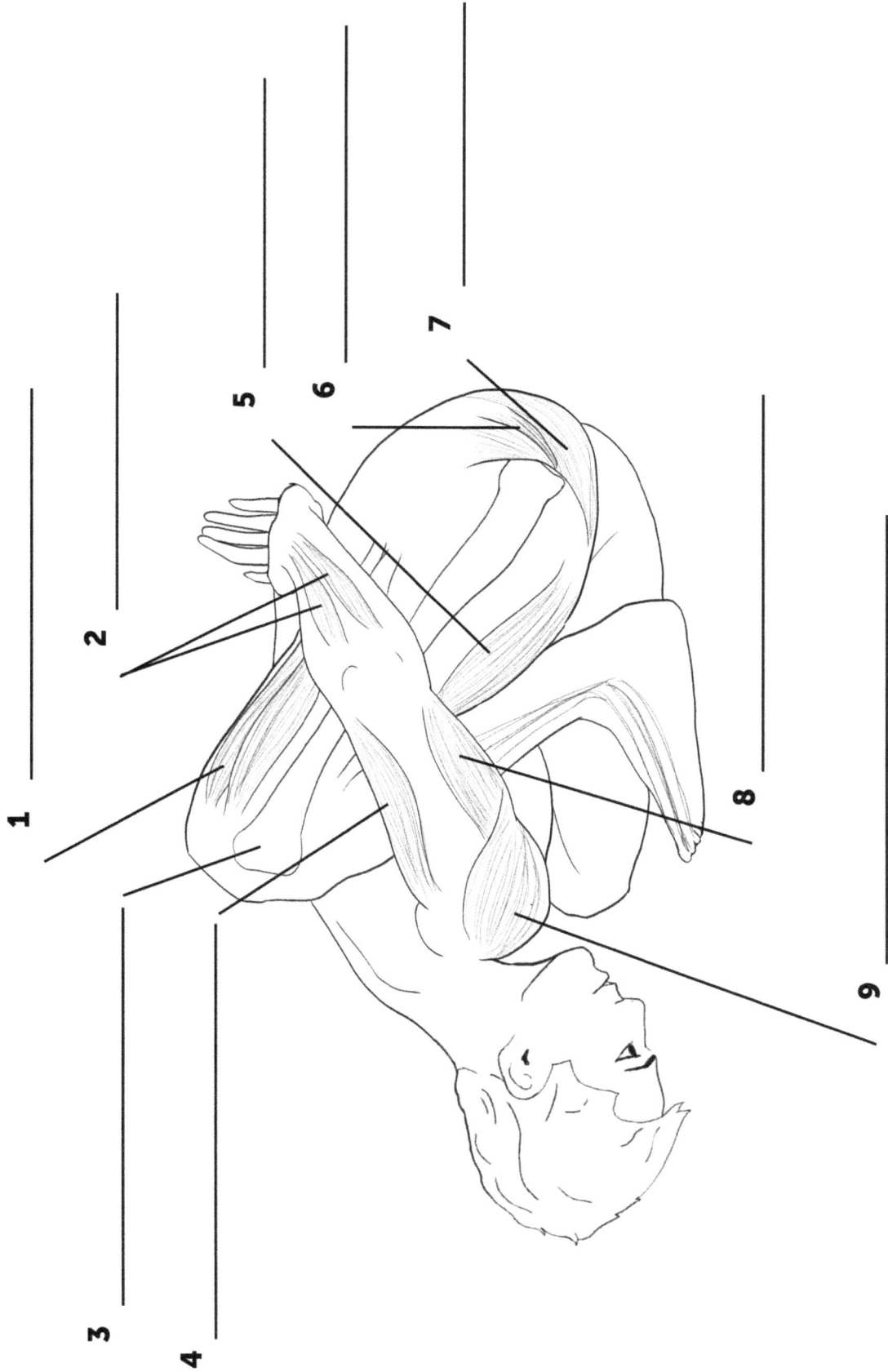

6. MARICHI'S POSE II

1. QUADRICEPS

2. PRONATORS

3. FEMUR

4. BICEPS BRACHII

5. HAMSTRINGS

6. PIRIFORMIS

7. GLUTEUS MAXIMUS

8. TRICEPS BRACHII

9. DELTOID

7. MARICHI'S POSE III

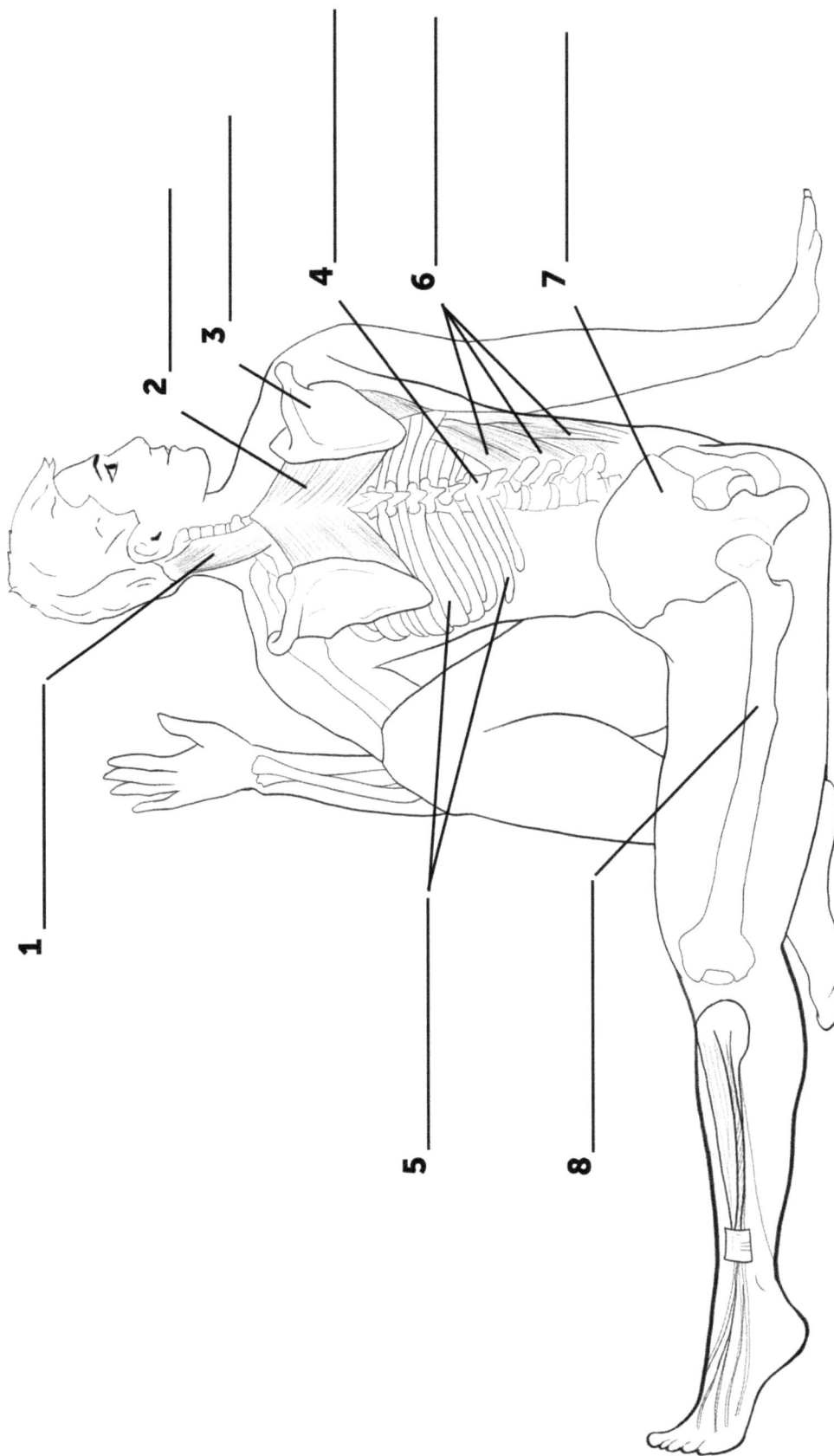

7. MARICHI'S POSE III

1. SPLENIUS CAPITIS

2. RHOMBOIDS

3. SCAPULA

4. SPINE

5. RIBS

6. ERECTOR SPINAE

7. PELVIS

8. FEMUR

8. PYRAMID POSE

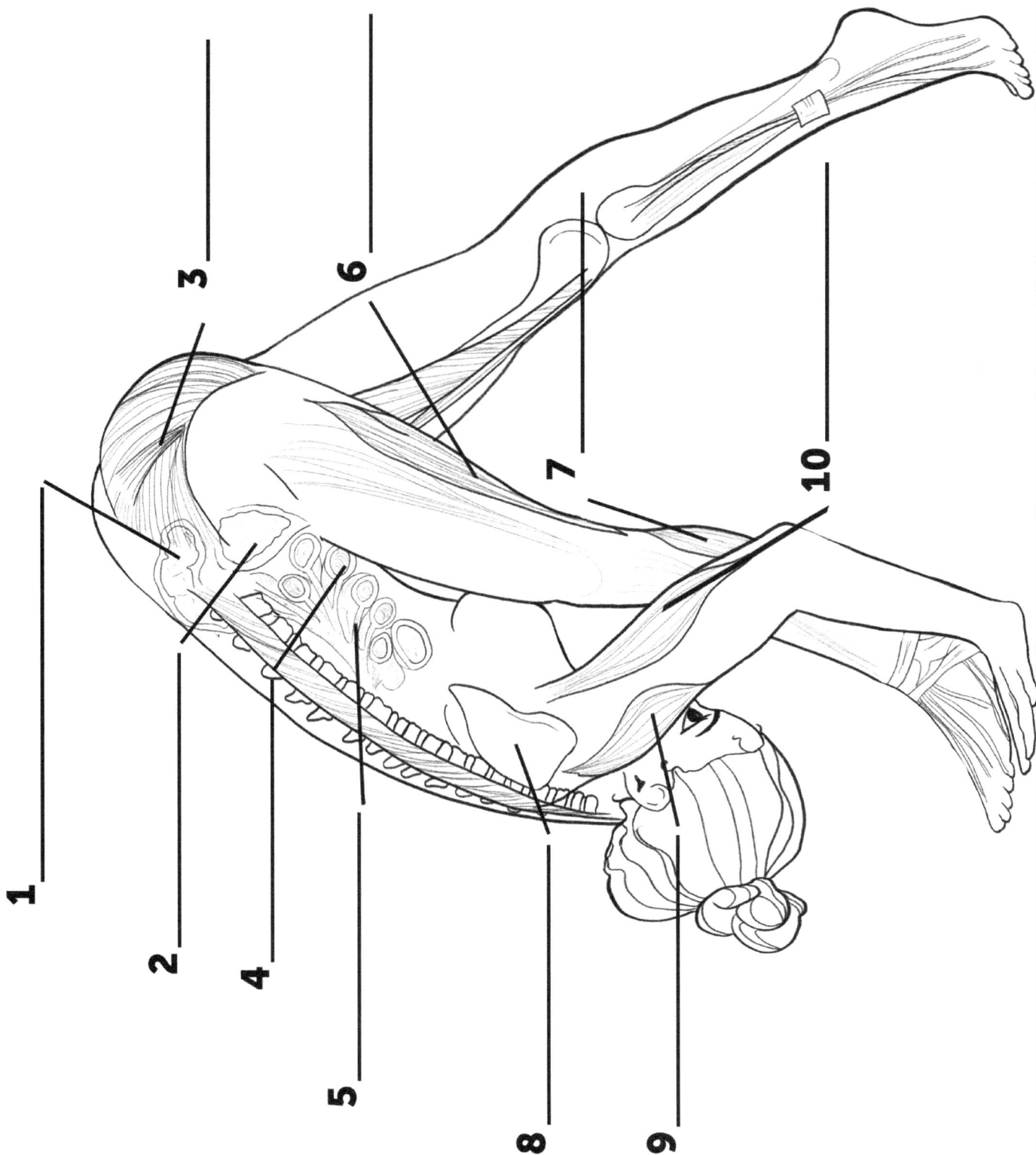

8. PYRAMID POSE

1.   RECTUM

2.   URINARY BLADDER

3.   PIRIFORMIS

4.   COILS OF SMALL INTESTINE

5.   MESENTERY OF SMALL INTESTINE

6.   HAMSTRINGS

7.   GASTROCNEMIUS

8.   SCAPULA

9.   DELTOID

10.   TRICEPS BRACHII

9. WARRIOR I POSE

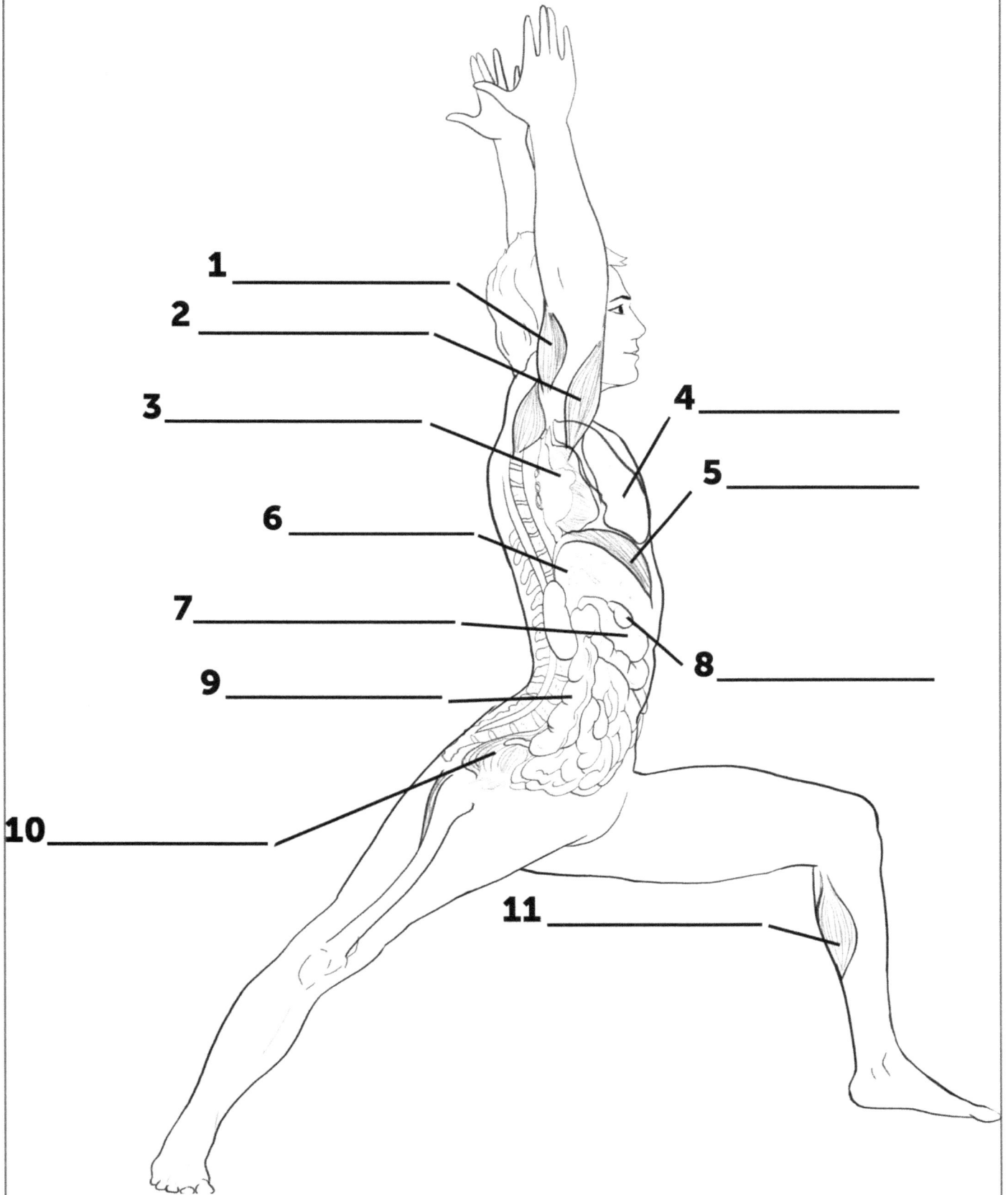

1 _____

2 _____

3 _____

4 _____

5 _____

6 _____

7 _____

8 _____

9 _____

10 _____

11 _____

9. WARRIOR I POSE

1. BICEPS BRACHII
2. TRICEPS BRACHII
3. HEART
4. LUNGS
5. DIAPHRAGM
6. LIVER
7. STOMACH
8. GALLBLADDER
9. ASCENDING COLON
10. RECTUM
11. GASTROCNEMIUS

10. TWISTED WARRIOR POSE

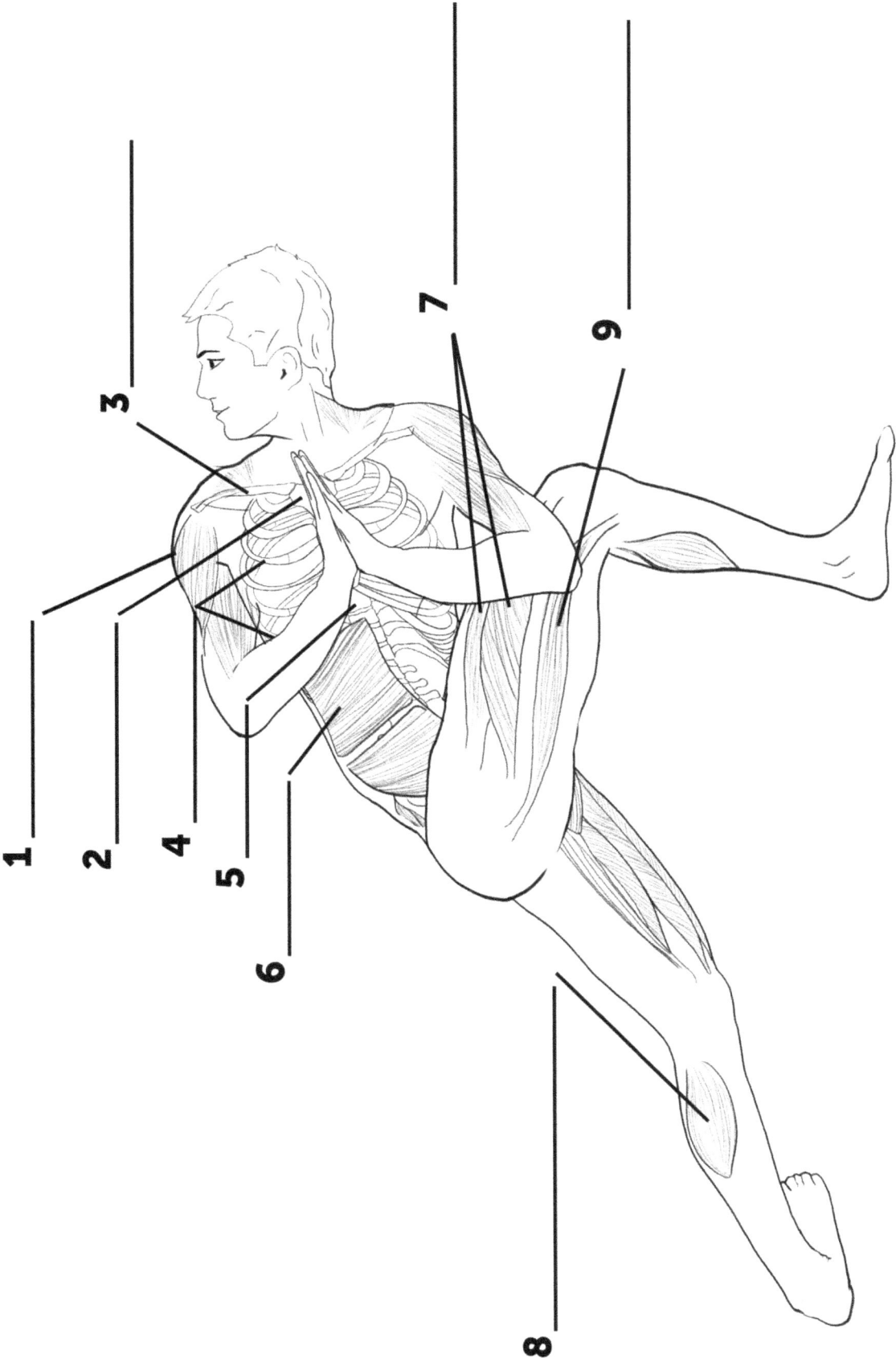

1

2

3

4

5

6

7

8

9

10. TWISTED WARRIOR POSE

1. DELTOID

2. STERNUM

3. COLLARBONE

4. RIBS

5. SPINE

6. INTERNAL OBLIQUE

7. QUADRICEPS

8. GASTROCNEMIUS

9. HAMSTRINGS

11. TWISTED TRIANGLE POSE

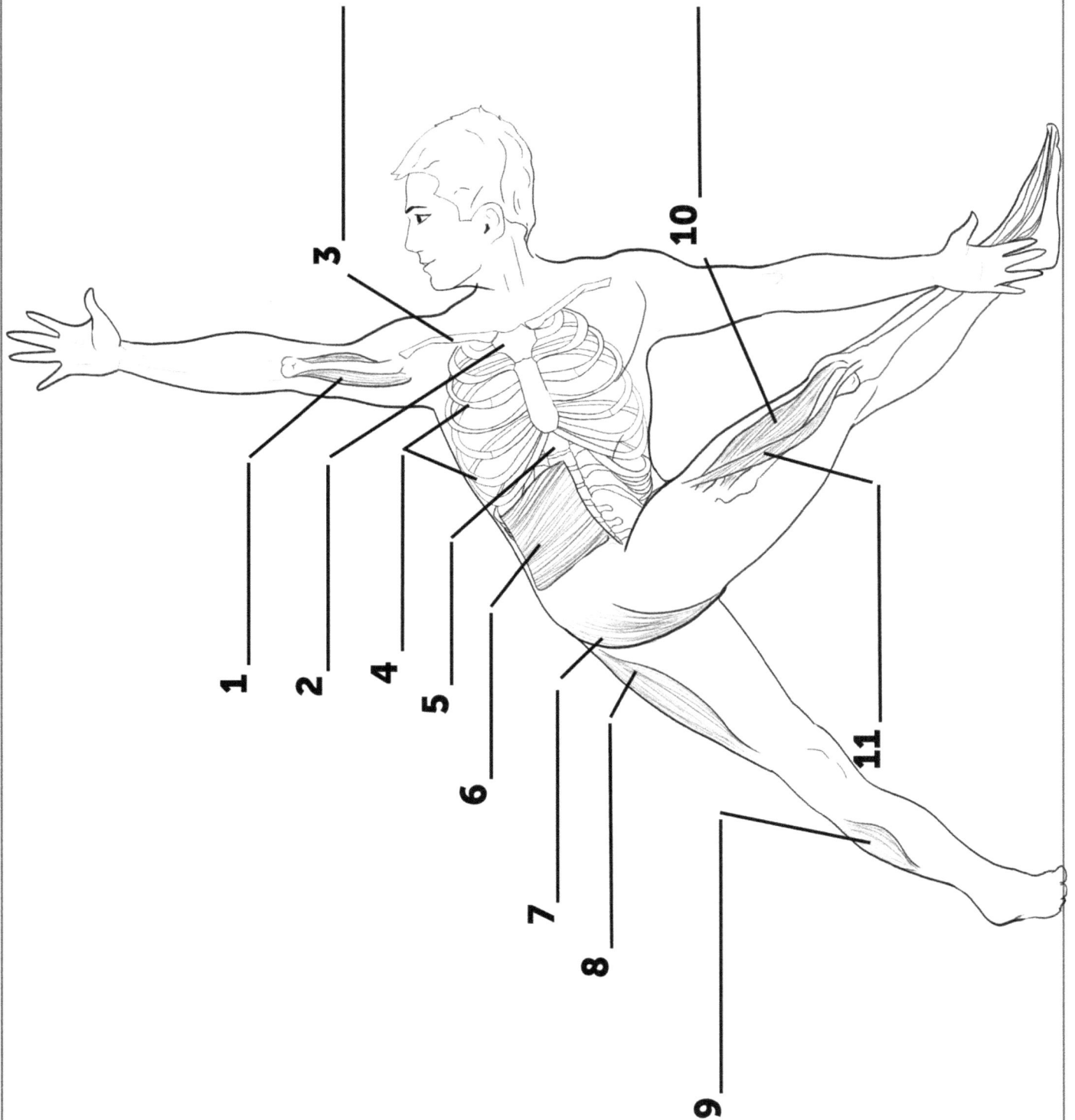

1

2

3

4

5

6

7

8

9

10

11

11. TWISTED TRIANGLE POSE

1. TRICEPS BRACHII
2. STERNUM
3. COLLARBONE
4. RIBS
5. SPINE
6. INTERNAL OBLIQUE
7. GLUTEUS MAXIMUS
8. HAMSTRINGS
9. GASTROCNEMIUS
10. QUADRICEPS
11. SARTORIUS

12. BOUND TWISTED SIDE ANGLE POSE

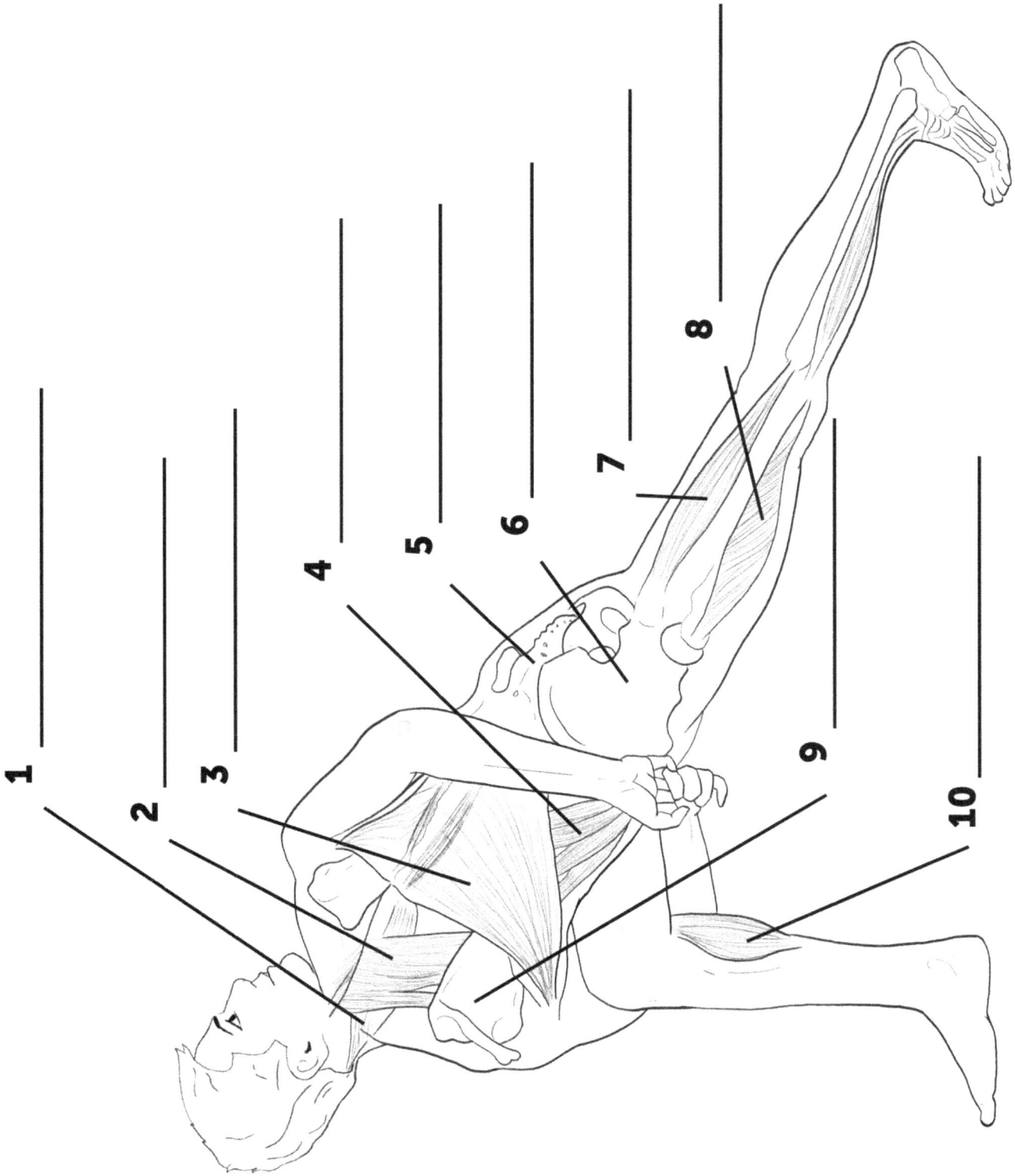

12. BOUND TWISTED SIDE ANGLE POSE

1. SPLENIUS CAPITIS

2. RHOMBOIDS

3. LATISSIMUS DORSI

4. ERECTOR SPINAE

5. SACRUM

6. PELVIS

7. HAMSTRINGS

8. QUADRICEPS

9. SCAPULA

10. GASTROCNEMIUS

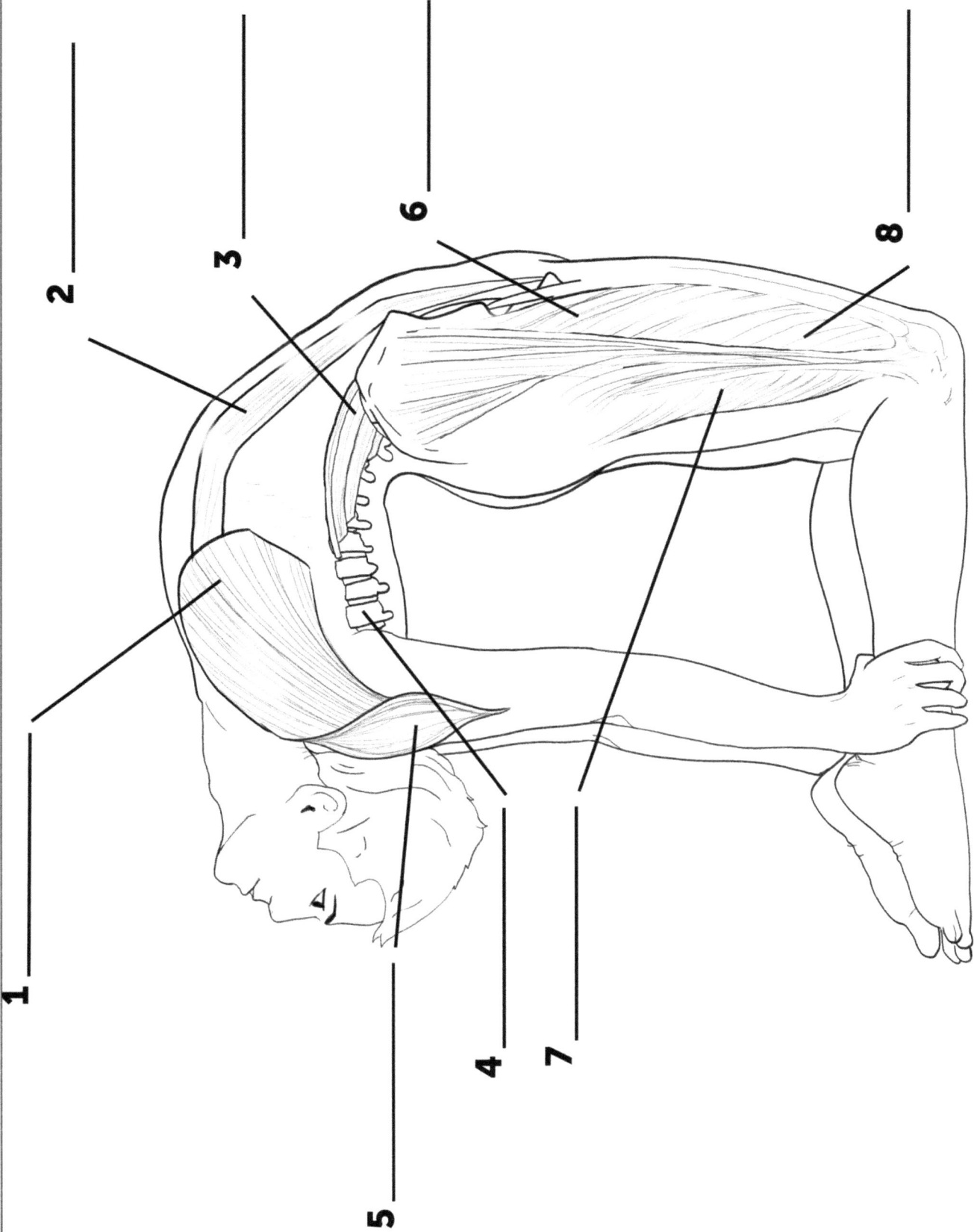

13. CAMEL POSE

1

2

3

6

8

4

7

5

13. CAMEL POSE

1. PECTORALIS MAJOR

2. RECTUS ABDOMINIS

3. PSOAS MAJOR

4. SPINE

5. DELTOID

6. RECTUS FEMORIS

7. HAMSTRINGS

8. VASTUS LATERALIS

14. WARRIOR II POSE

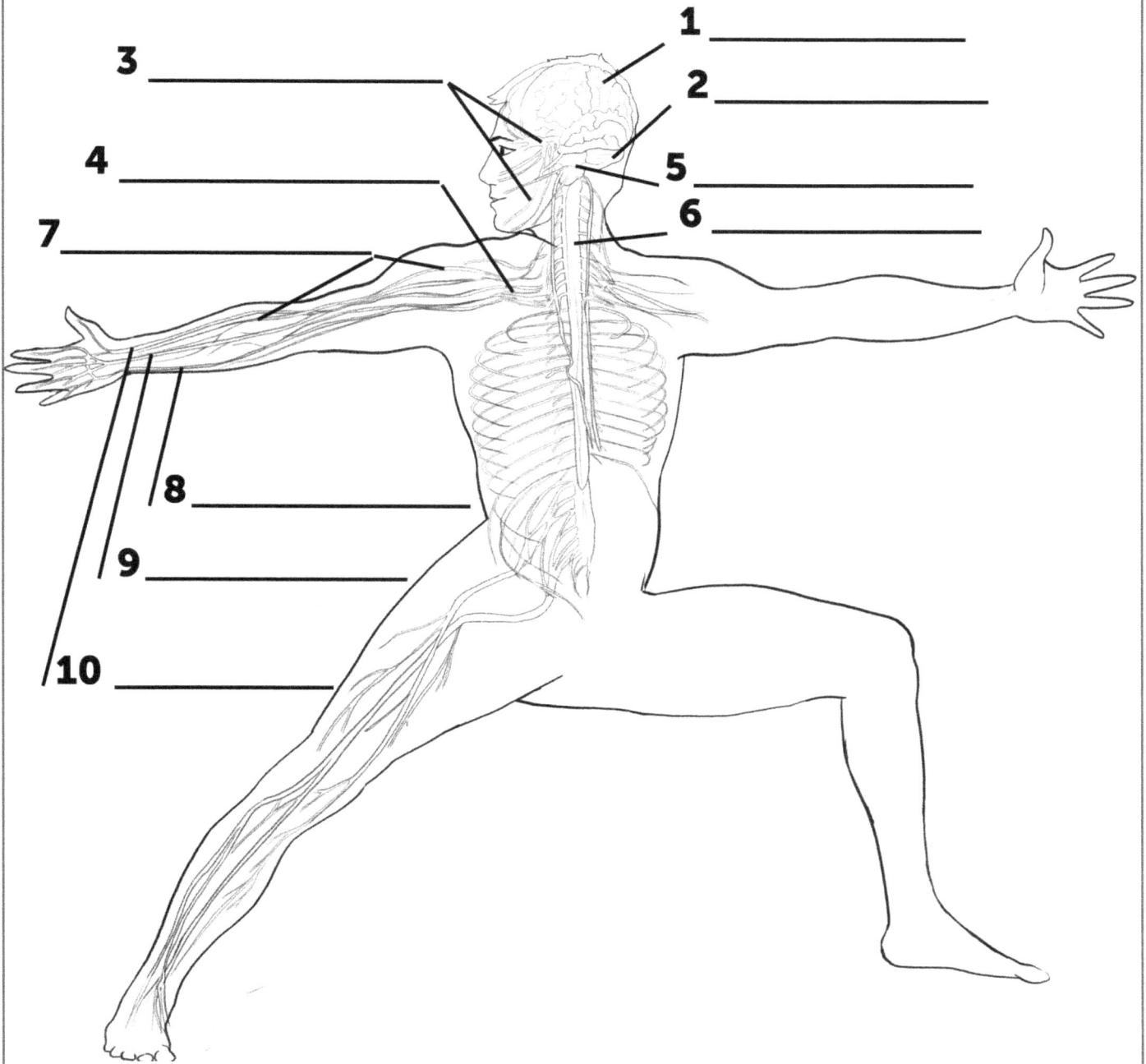

1 _____

3 _____

2 _____

4 _____

5 _____

7 _____

6 _____

8 _____

9 _____

10 _____

14. WARRIOR II POSE

1. CEREBRUM
2. CEREBELLUM
3. CRANIAL NERVES
4. BRACHIAL PLEXUS
5. BRAINSTEM
6. SPINAL CORD
7. MUSCULOCUTANEOUS
8. ULNAR
9. MEDIAN
10. RADIAL

15. WARRIOR III POSE

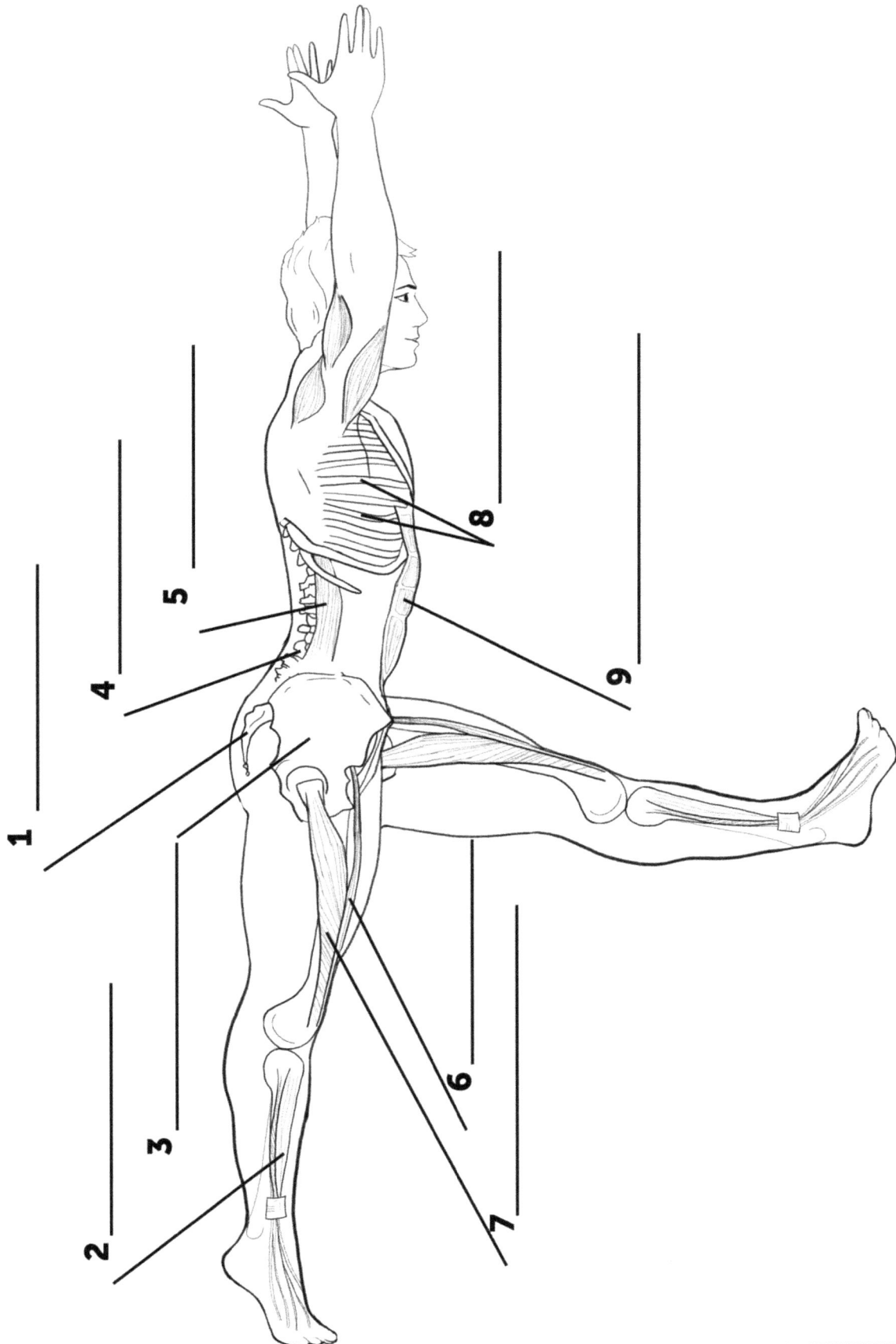

15. WARRIOR III POSE

1.   SACRUM

2.   TIBIALIS ANTERIOR

3.   PELVIS

4.   SPINE

5.   ERECTOR SPINAE

6.   SARTORIUS

7.   RECTUS FEMORIS

8.   RIBS

9.   RECTUS ABDOMINIS

16. REVERSE WARRIOR POSE

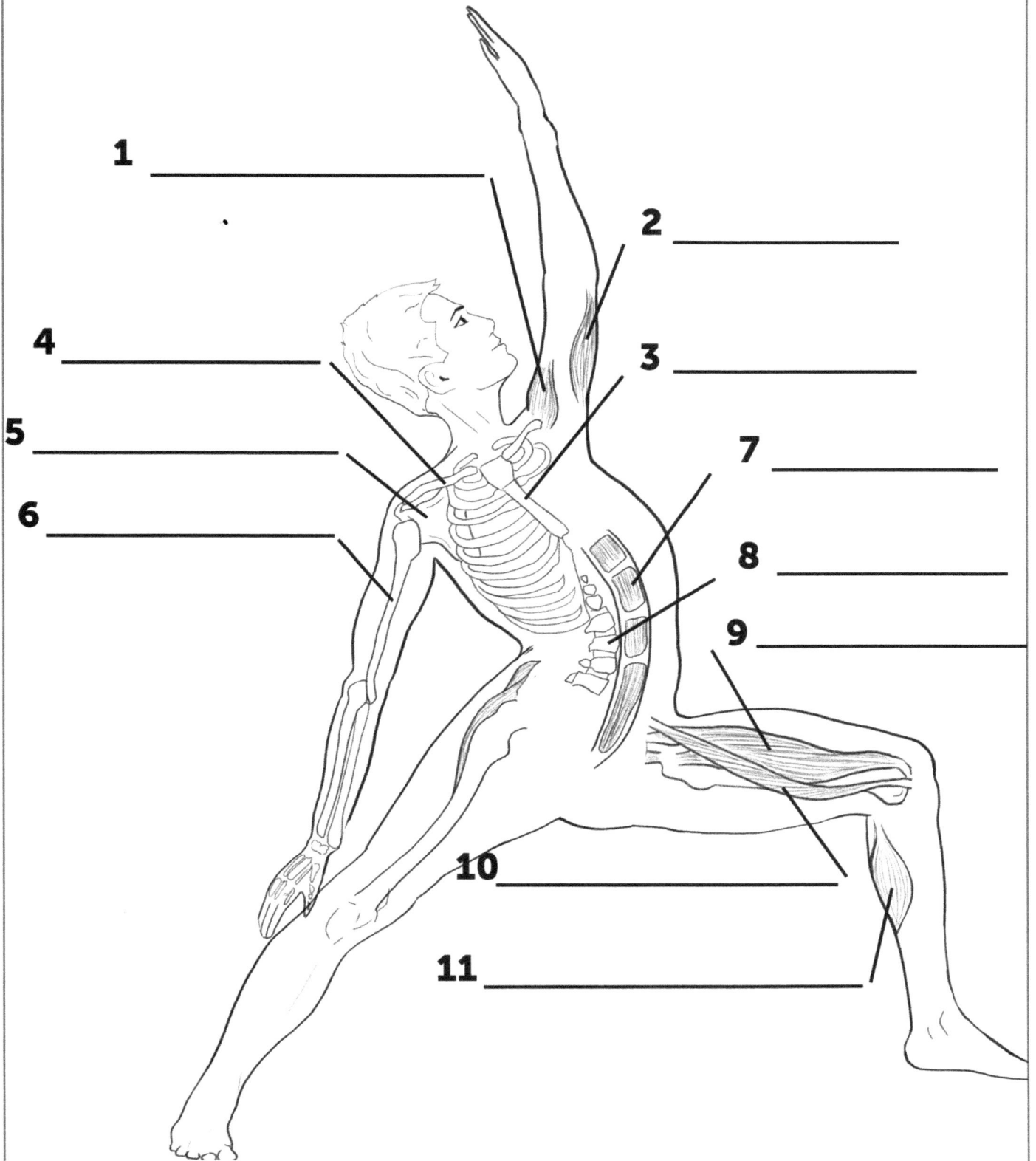

1 _____

2 _____

4 _____

3 _____

5 _____

7 _____

6 _____

8 _____

9 _____

10 _____

11 _____

16. REVERSE WARRIOR POSE

1. DELTOID
2. TRICEPS BRACHII
3. STERNUM
4. COLLARBONE
5. SCAPULA
6. HUMERUS
7. RECTUS ABDOMINIS
8. SPINE
9. RECTUS FEMORIS
10. SARTORIUS
11. GASTROCNEMIUS

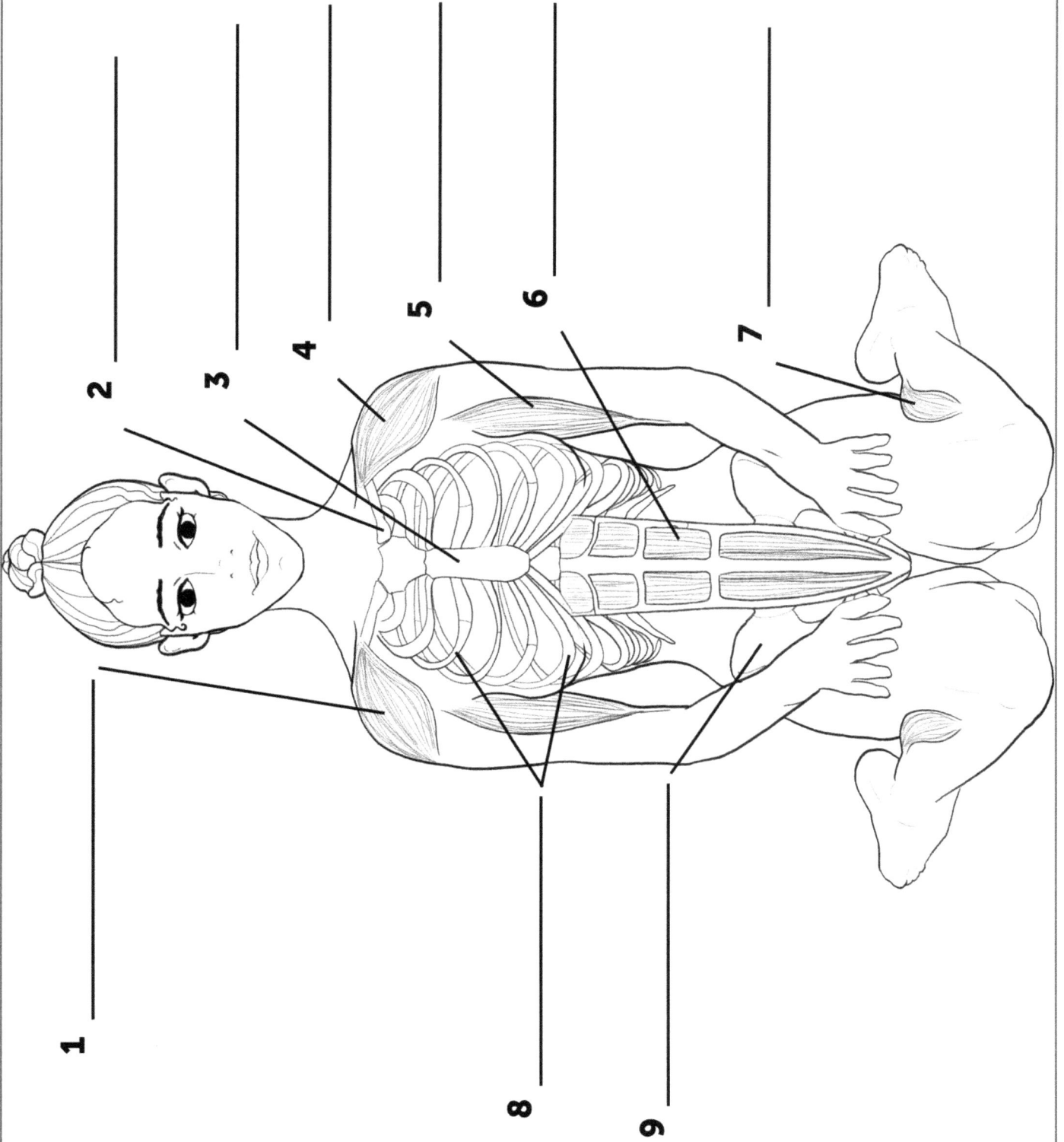

17. HERO POSE

1

2

3

4

5

6

7

8

9

17. HERO POSE

1. DELTOID

2. COLLARBONE

3. STERNUM

4. DELTOID

5. BICEPS BRACHII

6. RECTUS ABDOMINIS

7. GASTROCNEMIUS

8. RIBS

9. PELVIS

18. HALF RECLINED HERO

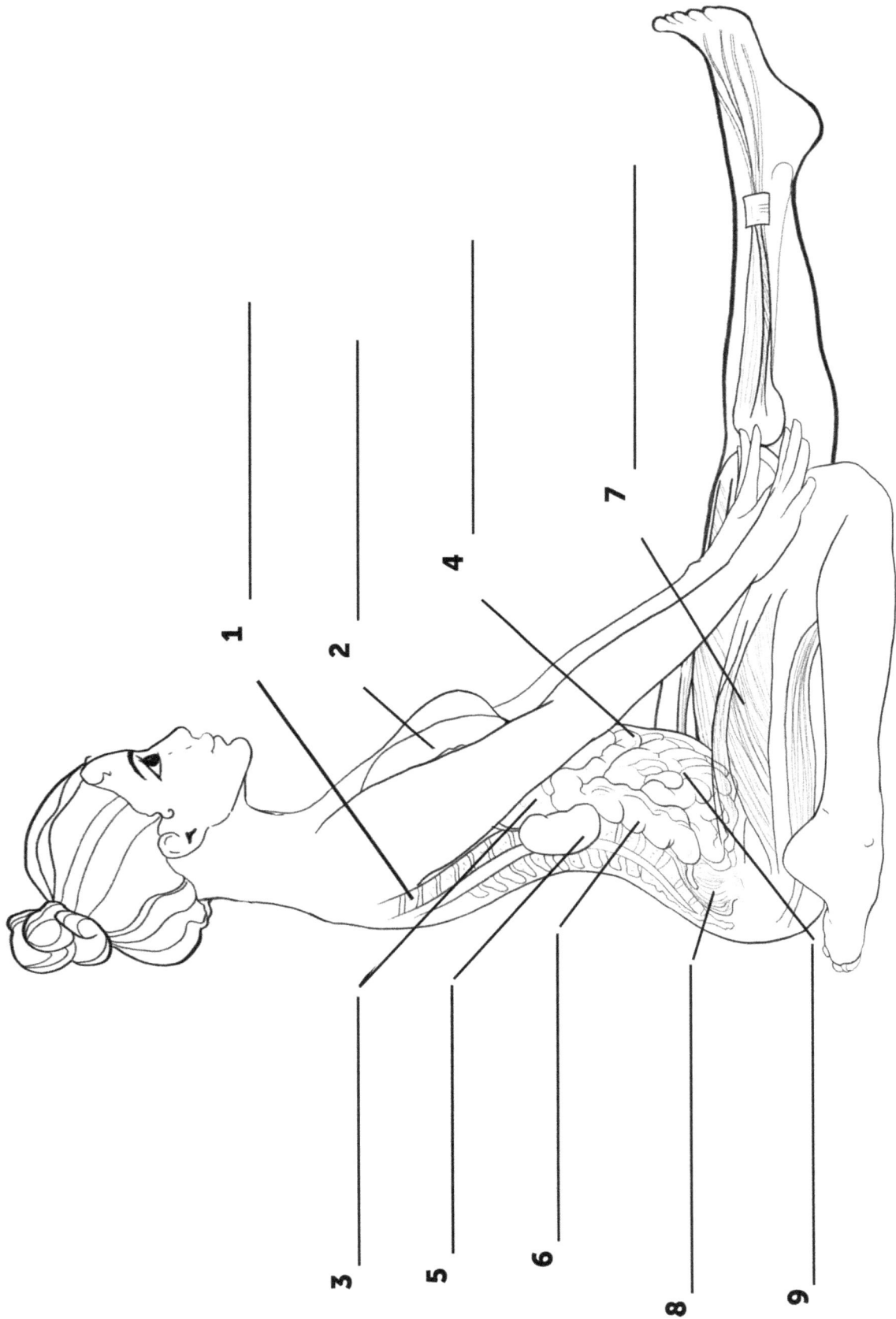

1

2

4

7

3

5

6

8

9

18. HALF RECLINED HERO

1. SPINE
2. LUNGS
3. LIVER
4. TRANSVERSE COLON
5. KIDNEY
6. ASCENDING COLON
7. QUADRICEPS
8. RECTUM
9. COILS OF SMALL INTESTINE

19. RECLINING HERO POSE

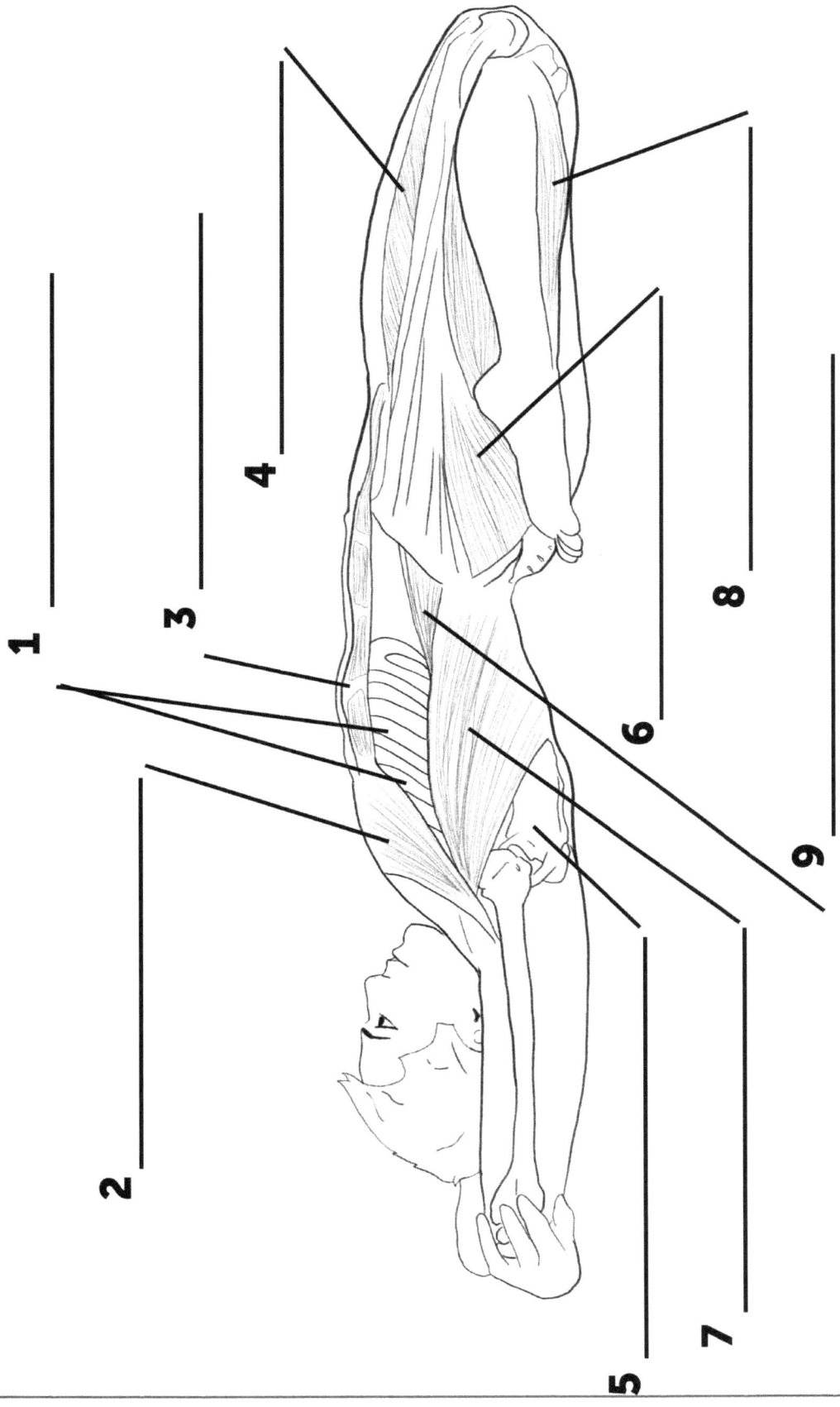

1

2

3

4

5

6

7

8

9

19. RECLINING HERO POSE

1. RIBS
2. PECTORALIS MAJOR
3. RECTUS ABDOMINIS
4. VASTUS LATERALIS
5. SCAPULA
6. GLUTEUS MAXIMUS
7. LATISSIMUS DORSI
8. TIBIALIS ANTERIOR
9. PSOAS MAJOR

20. EXTENDED HAND-TO-BIG-TOE POSE

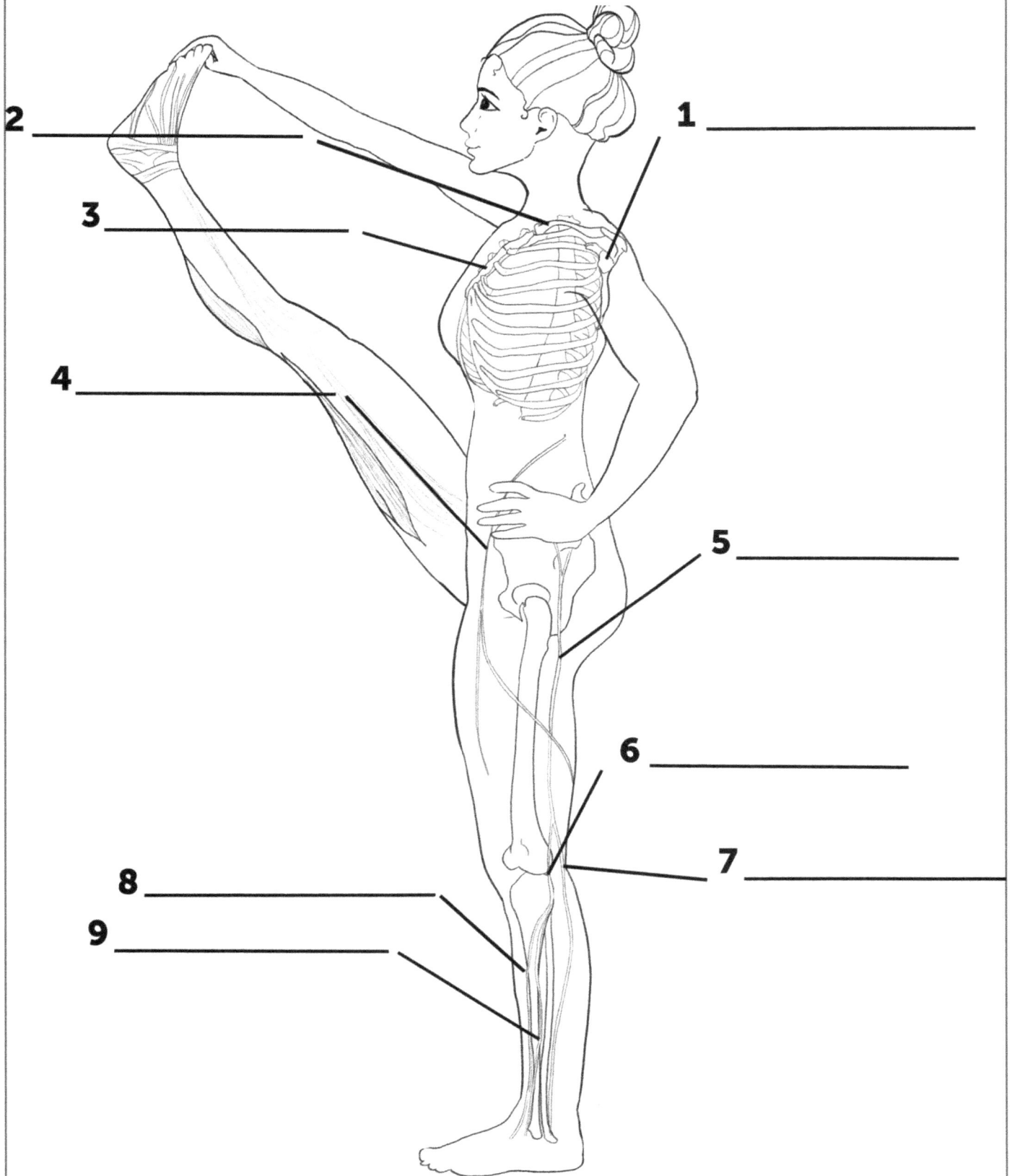

2 _____

3 _____

4 _____

1 _____

5 _____

6 _____

7 _____

8 _____

9 _____

20. EXTENDED HAND-TO-BIG-TOE POSE

1. SCAPULA

2. COLLARBONE

3. STERNUM

4. LATERAL FEMORAL CUTANEOUS NERVE

5. SCIATIC NERVE

6. COMMON PERONEAL NERVE

7. TIBIAL NERVE

8. DEEP PERONEAL NERVE

9. SUPERFICIAL PERONEAL NERVE

21. PIGEON POSE

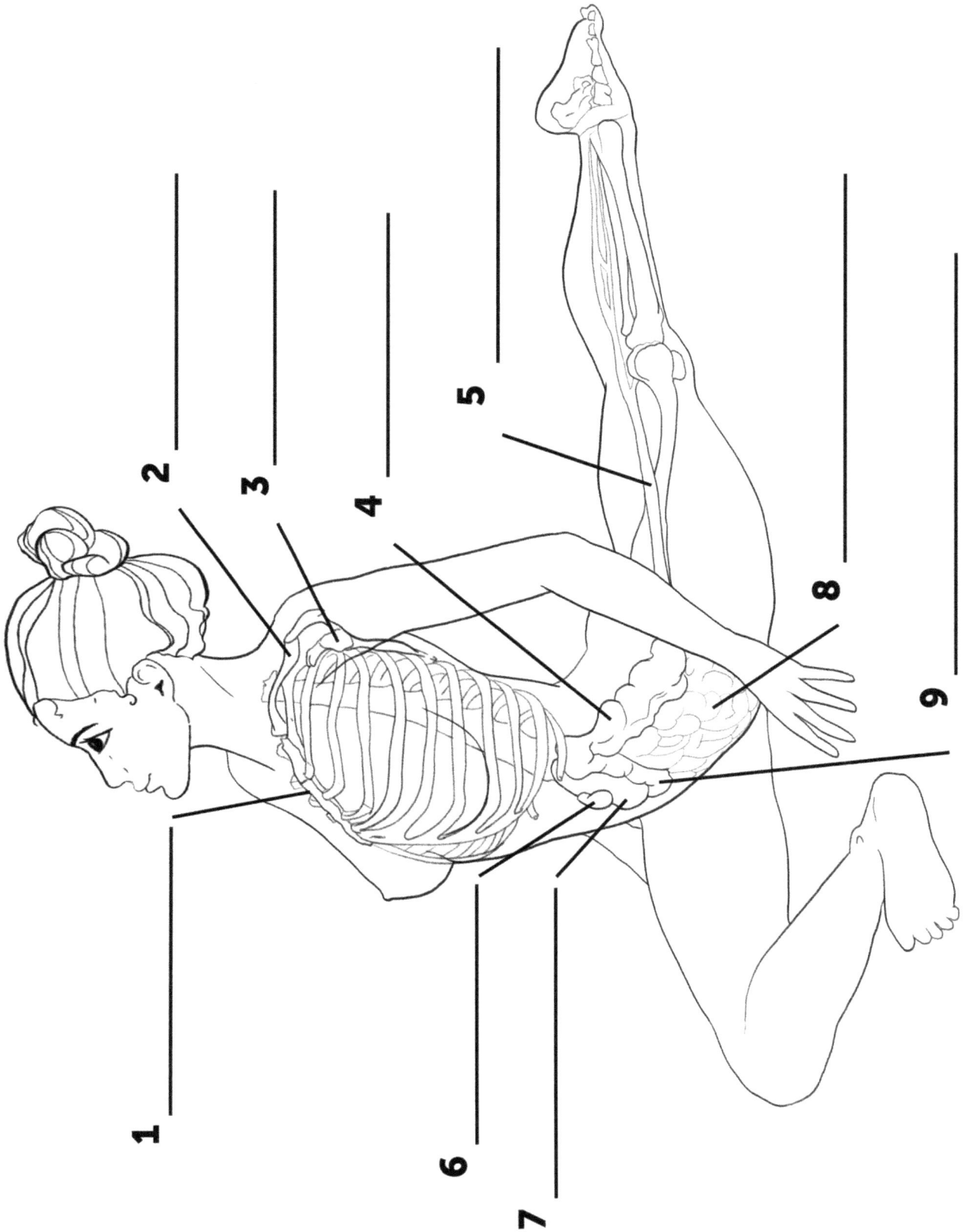

1

2

3

4

5

6

7

8

9

21. PIGEON POSE

1. STERNUM

2. COLLARBONE

3. SCAPULA

4. ASCENDING COLON

5. SCIATIC NERVE

6. GALLBLADDER

7. STOMACH

8. COILS OF SMALL INTESTINE

9. TRANSVERSE COLON

22. THREAD THE NEEDLE

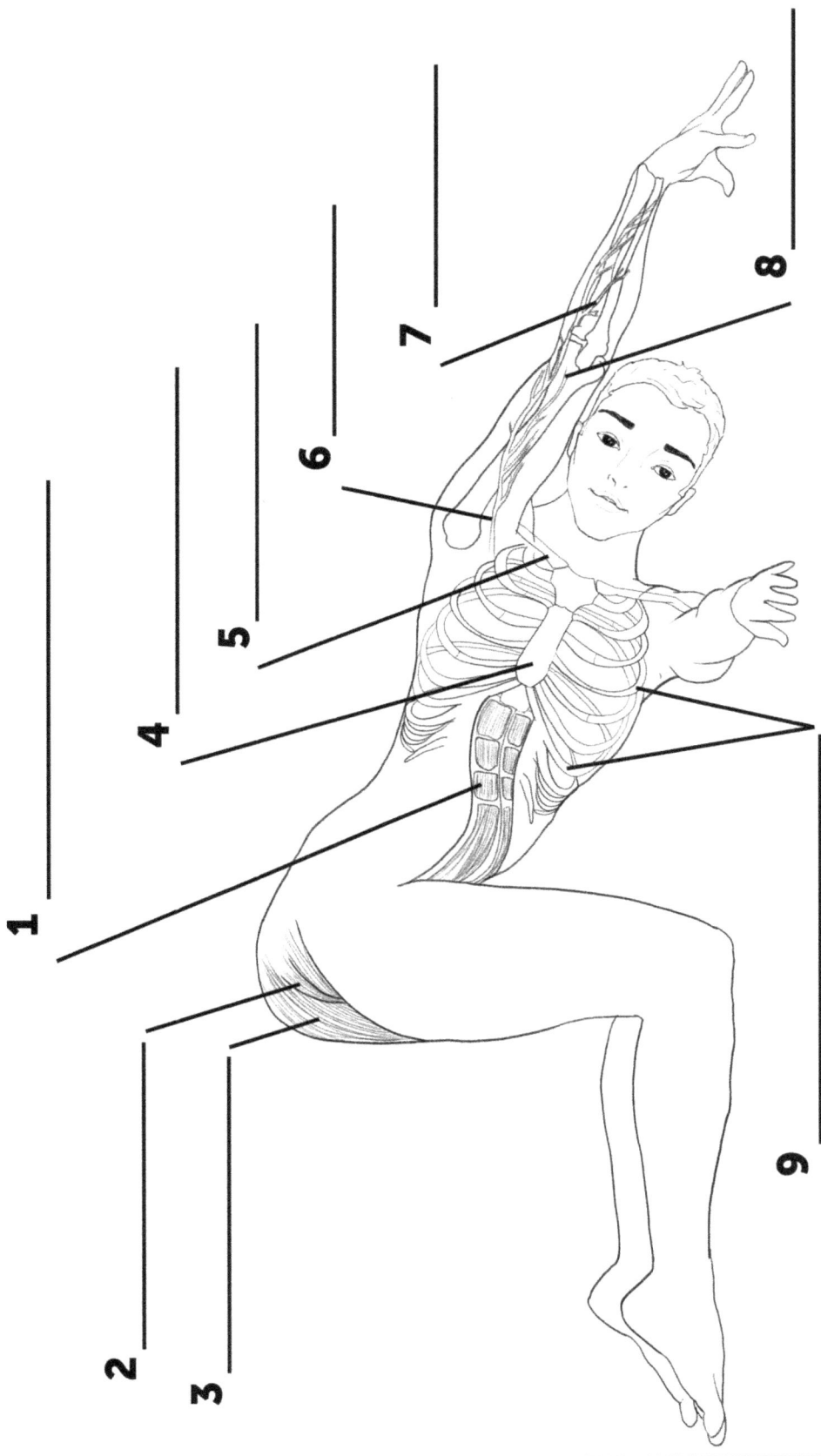

1

2

3

4

5

6

7

8

9

22. THREAD THE NEEDLE

1. RECTUS ABDOMINIS

2. PIRIFORMIS

3. GLUTEUS MAXIMUS

4. STERNUM

5. COLLARBONE

6. RADIAL NERVE

7. POSTERIOR INTEROSSEOUS NERVE

8. ANCONEUS

9. RIBS

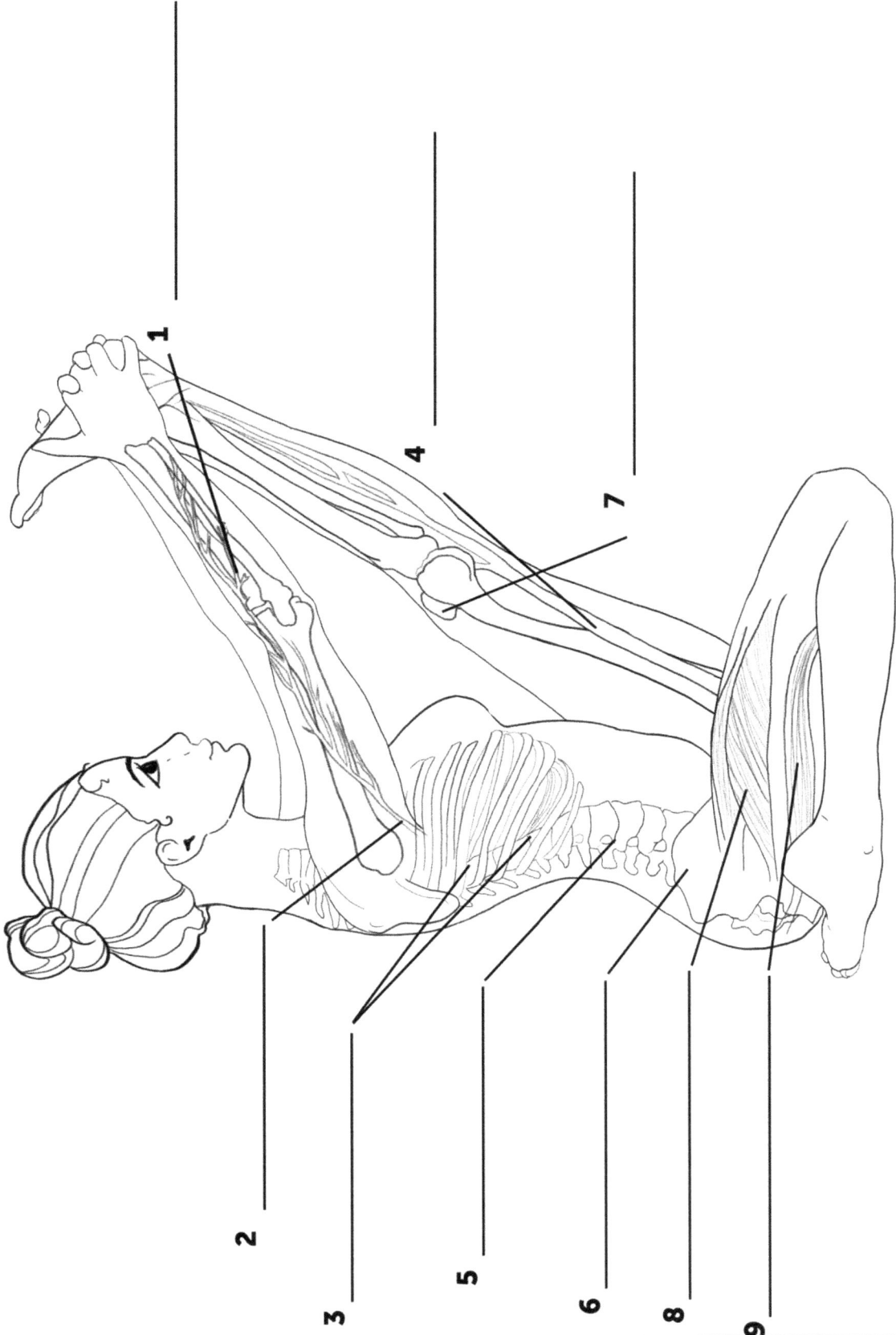

23. HERON POSE

1

2

3

4

5

6

7

8

9

23. HERON POSE

1. POSTERIOR INTEROSSEOUS NERVE

2. RADIAL NERVE

3. RIBS

4. SCIATIC NERVE

5. SPINE

6. PELVIS

7. PATELLA

8. QUADRICEPS

9. HAMSTRINGS

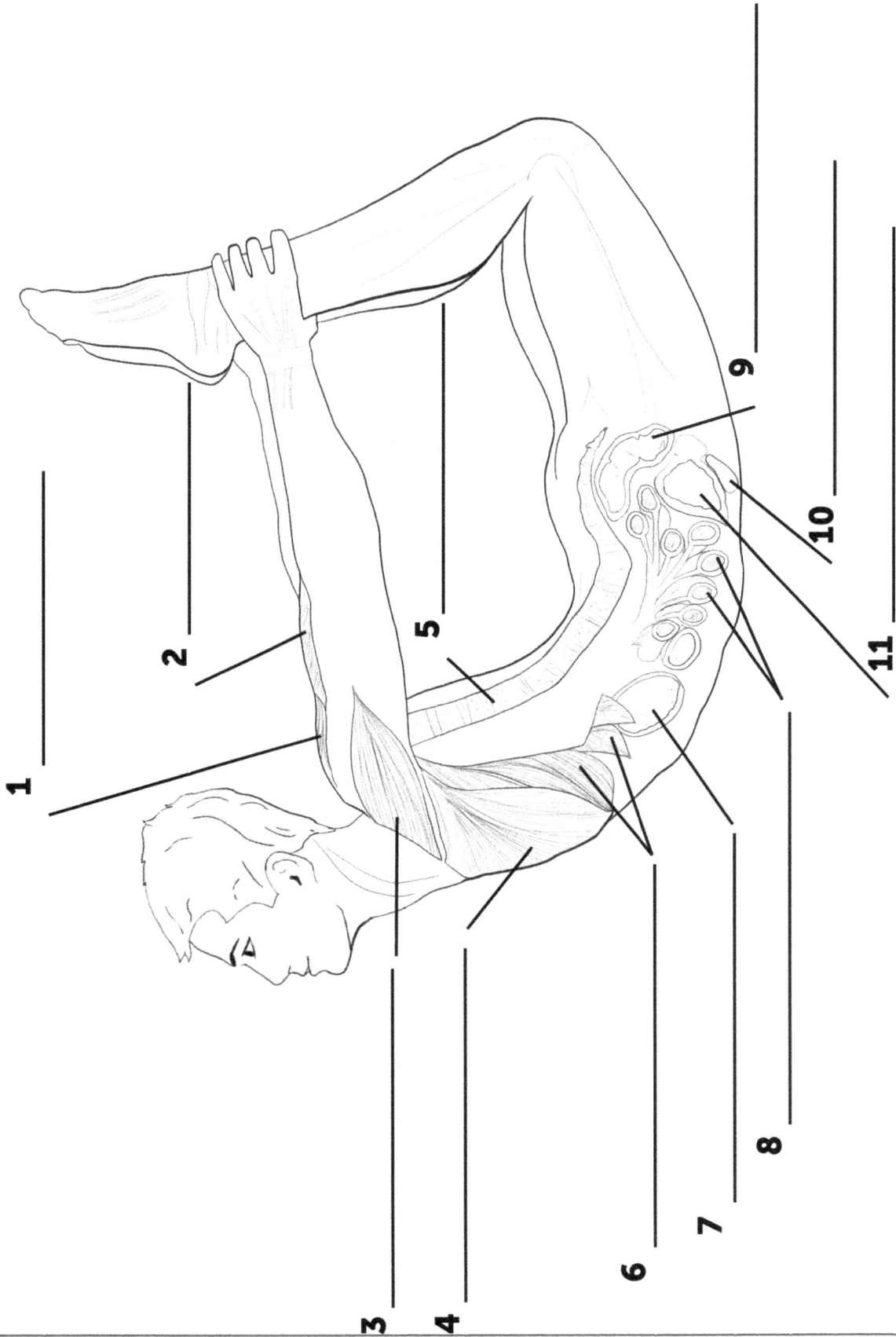

24. BOW POSE

24. BOW POSE

1. POSTERIOR DELTOID
2. TRICEPS BRACHII
3. ANTERIOR DELTOID
4. PECTORALIS MAJOR
5. SPINE
6. SERRATUS ANTERIOR
7. STOMACH
8. COILS OF SMALL INTESTINE
9. RECTUM
10. PUBIC BONE
11. URINARY BLADDER

25. UPWARD BOW OR WHEEL POSE

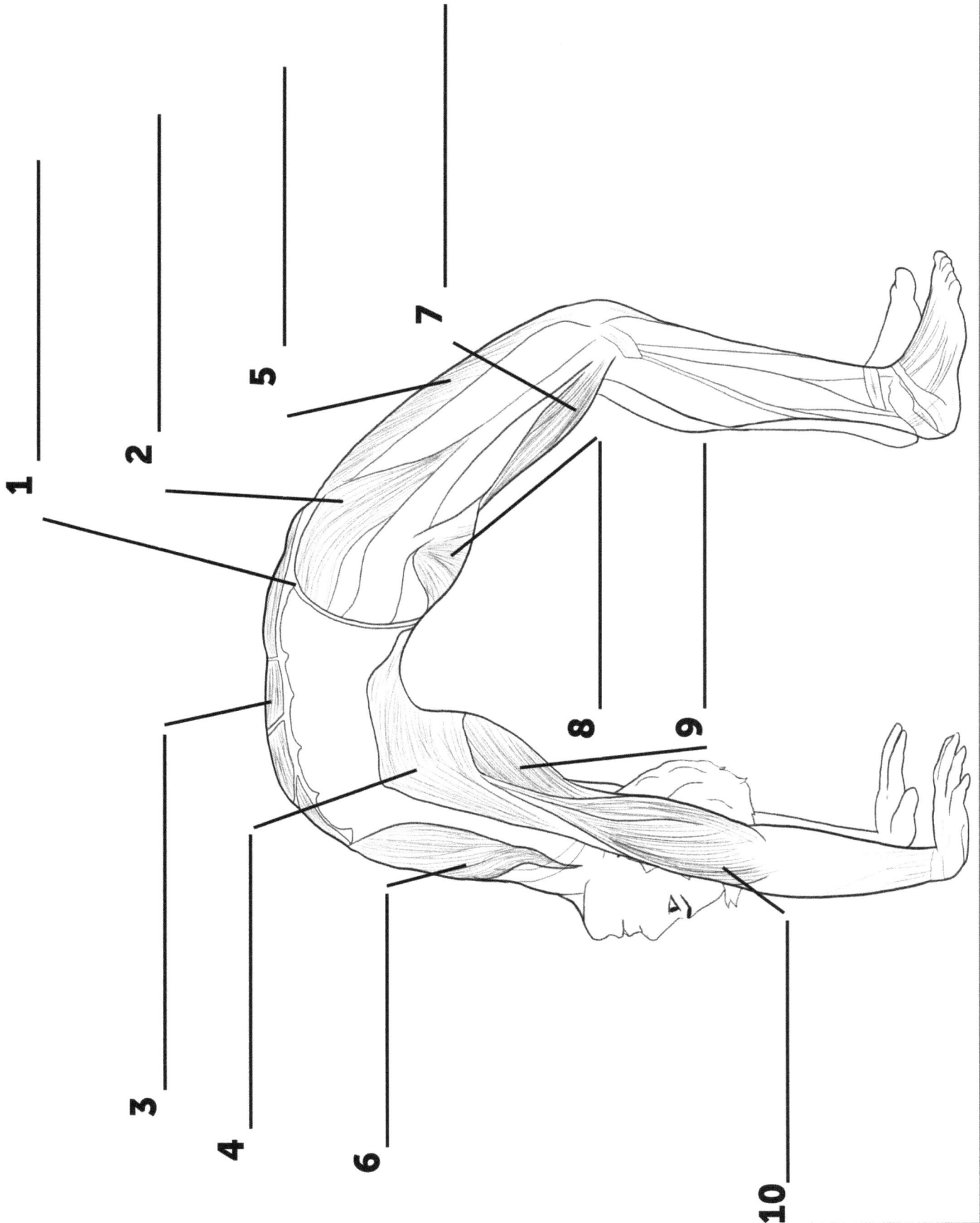

1

2

3

4

5

6

7

8

9

10

25. UPWARD BOW OR WHEEL POSE

1. ILIOPSOAS
2. TENSOR FASCIA LATA
3. RECTUS ABDOMINIS
4. LATISSIMUS DORSI
5. QUADRICEPS
6. PECTORALIS MAJOR
7. HAMSTRINGS
8. GLUTEUS MAXIMUS
9. ERECTOR SPINAE
10. TRICEPS BRACHII

26. LIZARD POSE

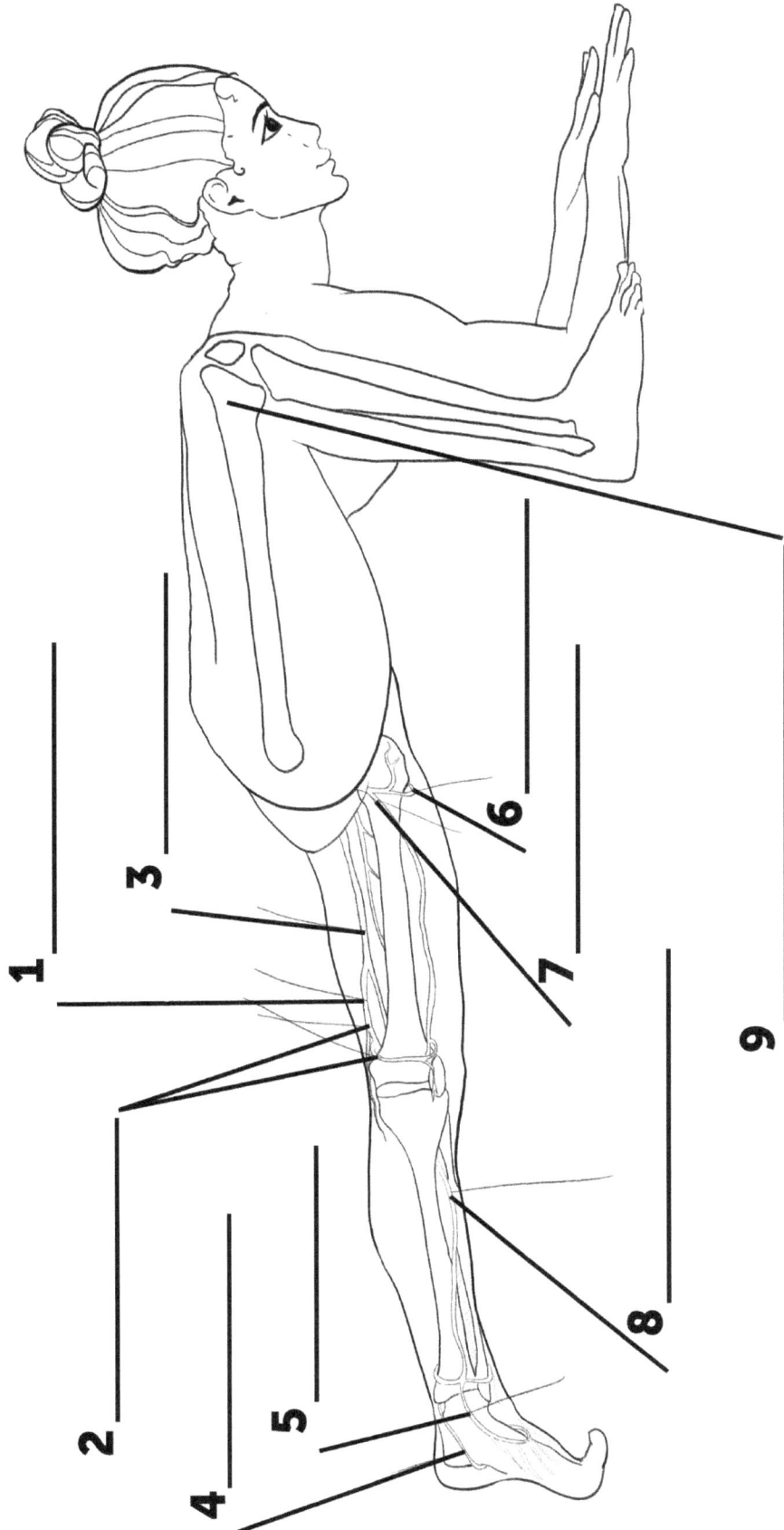

1

2

3

4

5

6

7

8

9

26. LIZARD POSE

1. ADDUCTOR HIATUS
2. GENICULAR ARTERIES
3. FEMORAL ARTERY
4. MEDIAL PLANTAR ARTERY
5. DORSALIS PEDIS ARTERY
6. LATERAL CIRCUMFLEX FEMORAL ARTERY
7. DESCENDING BRANCH
8. ANTERIOR TIBIAL ARTERY
9. FEMUR

27. ONE-LEGGED KING PIGEON POSE

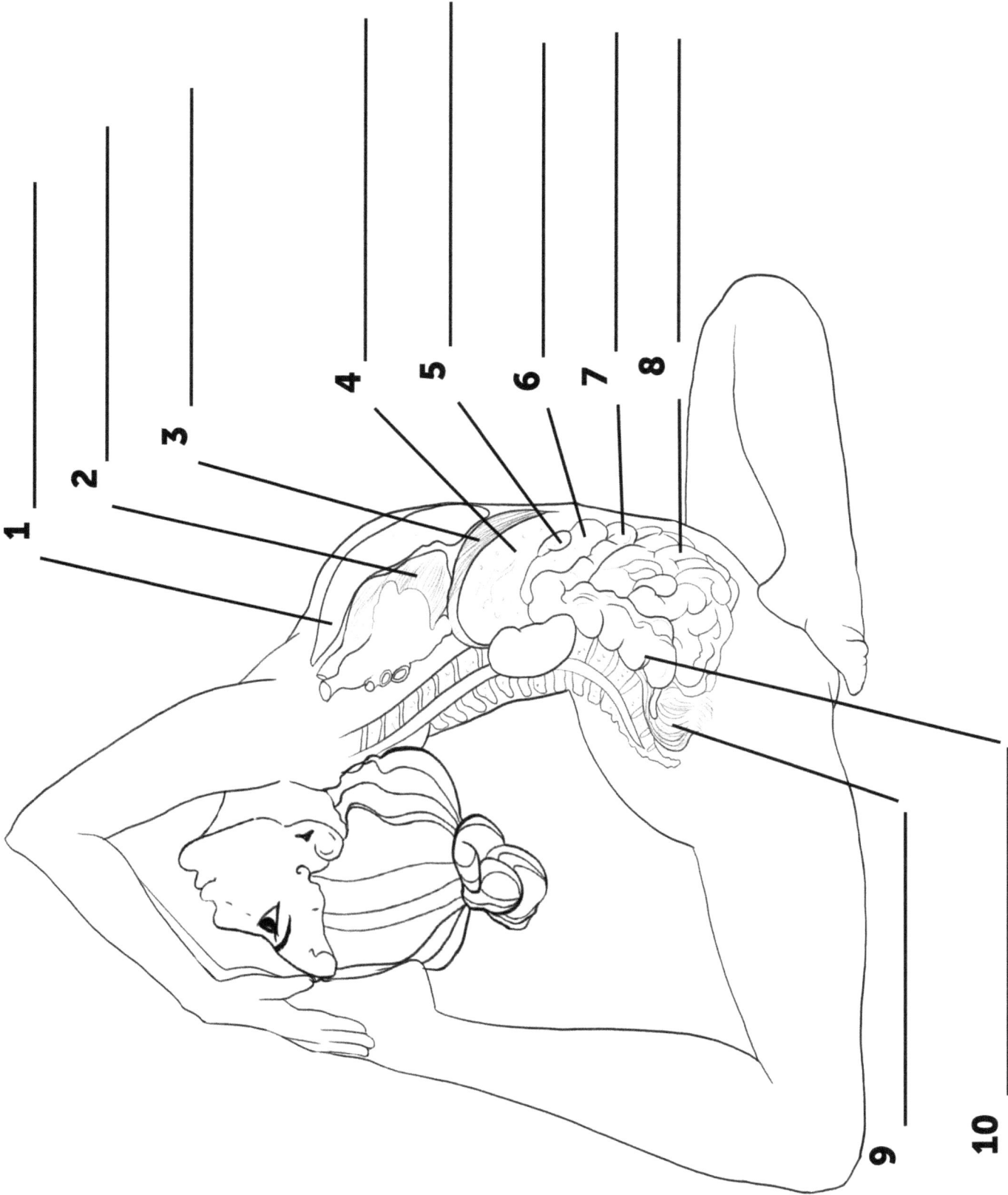

1

2

3

4

5

6

7

8

9

10

27. ONE-LEGGED KING PIGEON POSE

1. LUNGS

2. HEART

3. DIAPHRAGM

4. LIVER

5. GALLBLADDER

6. STOMACH

7. TRANSVERSE COLON

8. COILS OF SMALL INTESTINE

9. RECTUM

10. ASCENDING COLON

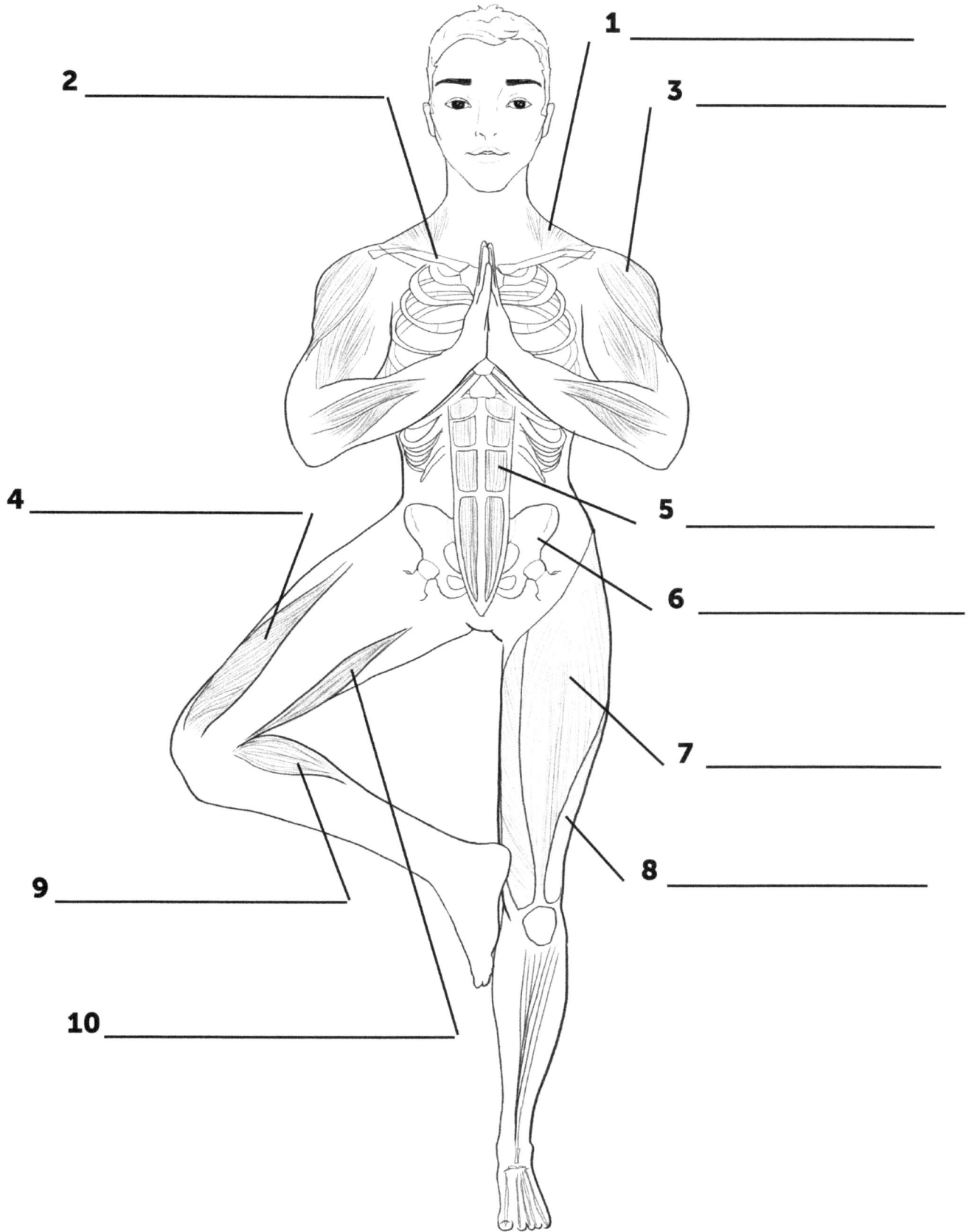

28. TREE POSE

1 _____

2 _____

3 _____

4 _____

5 _____

6 _____

7 _____

8 _____

9 _____

10 _____

28. TREE POSE

1. TRAPEZIUS
2. COLLARBONE
3. DELTOID
4. QUADRICEPS
5. RECTUS ABDOMINIS
6. PELVIS
7. RECTUS FEMORIS
8. VASTUS LATERALIS
9. GASTROCNEMIUS
10. HAMSTRINGS

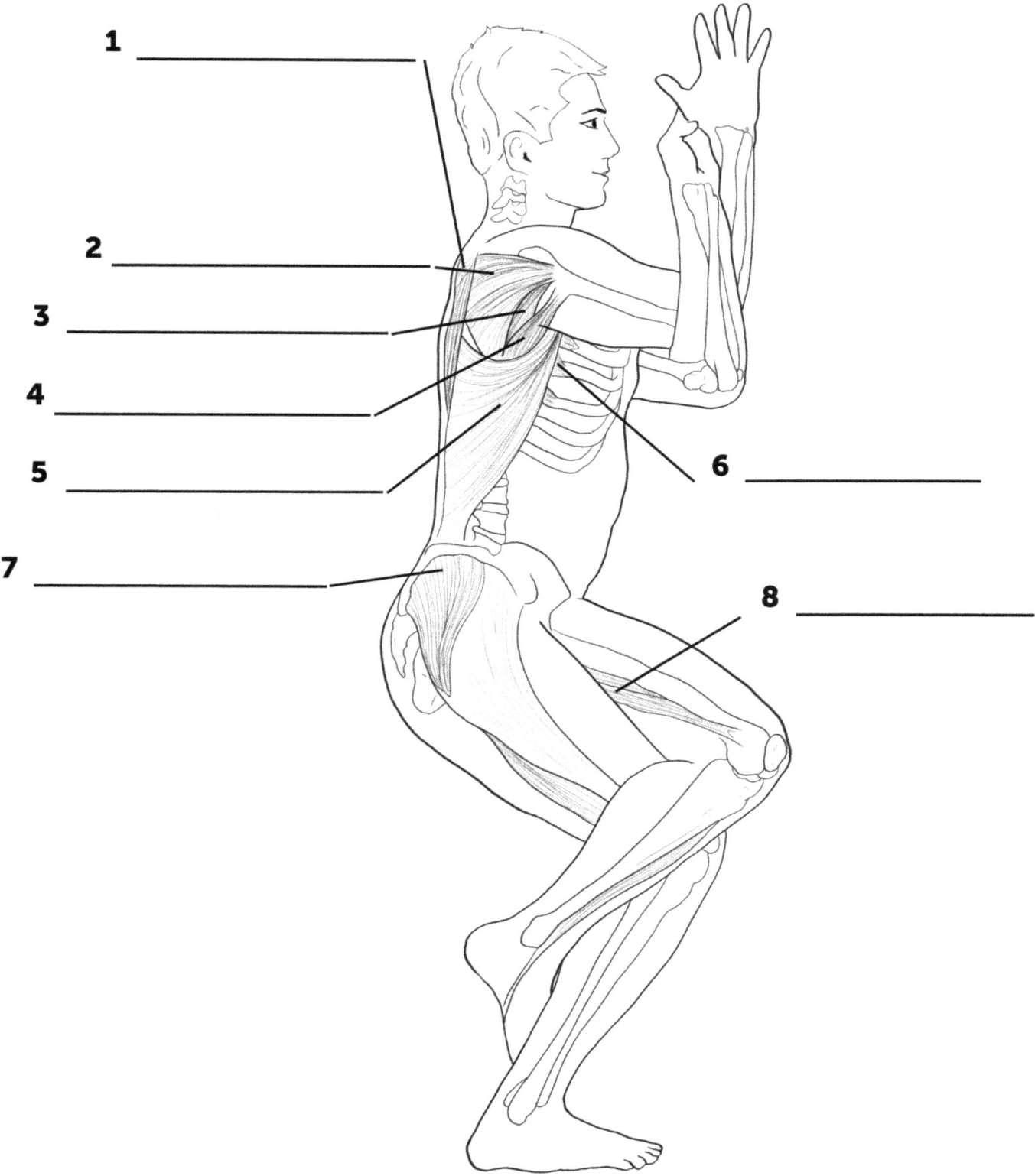

29. EAGLE POSE

1 _____

2 _____

3 _____

4 _____

5 _____

6 _____

7 _____

8 _____

29. EAGLE POSE

1. TRAPEZIUS
2. INFRASPINATUS
3. TERES MINOR
4. TERES MAJOR
5. LATISSIMUS DORSI
6. SERRATUS ANTERIOR
7. GLUTEUS MEDIUS
8. ADDUCTOR MAGNUS

30. HEAD TO KNEE POSE

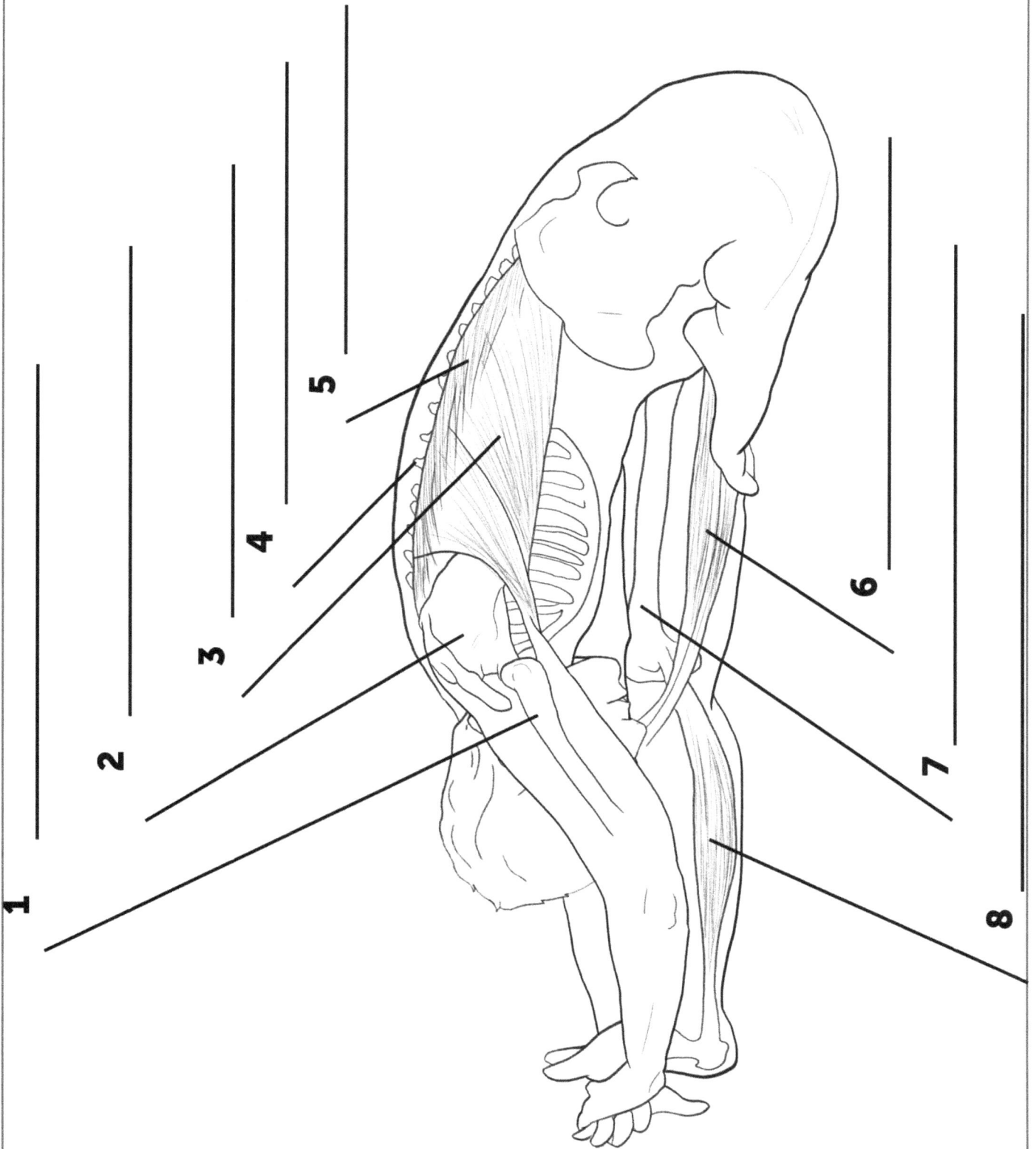

1

2

3

4

5

6

7

8

30. HEAD TO KNEE POSE

1.   HUMERUS

2.   SCAPULA

3.   LATISSIMUS DORSI

4.   SPINE

5.   ERECTOR SPINAE

6.   HAMSTRINGS

7.   FEMUR

8.   GASTROCNEMIUS

31. LORD OF THE DANCE POSE

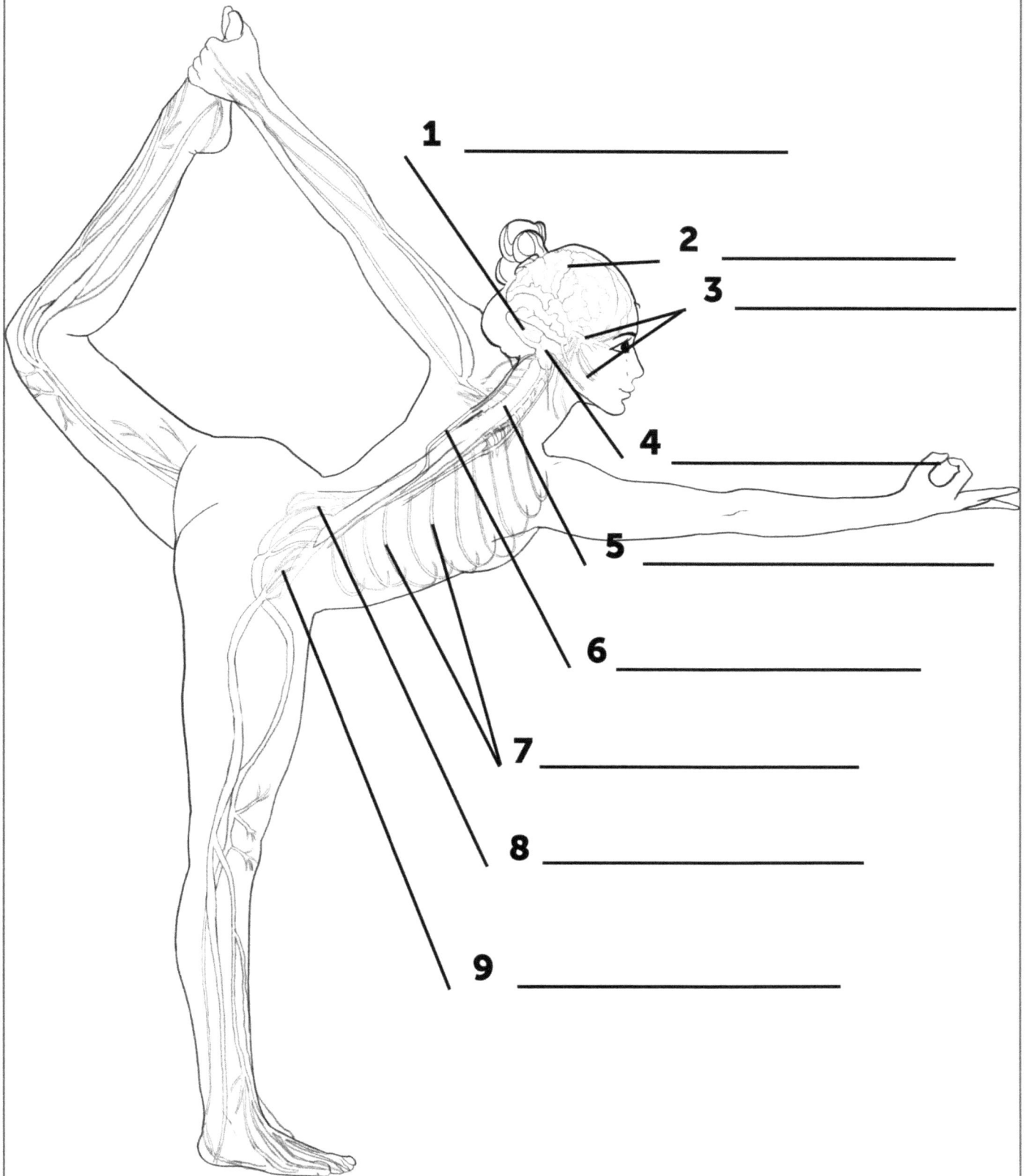

1 _____

2 _____

3 _____

4 _____

5 _____

6 _____

7 _____

8 _____

9 _____

31. LORD OF THE DANCE POSE

1. CEREBELLUM
2. CEREBRUM
3. CRANIAL NERVES
4. BRAINSTEM
5. SPINAL CORD
6. VAGUS
7. INTERCOSTALS
8. LUMBAR PLEXUS
9. SACRAL PLEXUS

32. TWIST CHAIR POSE

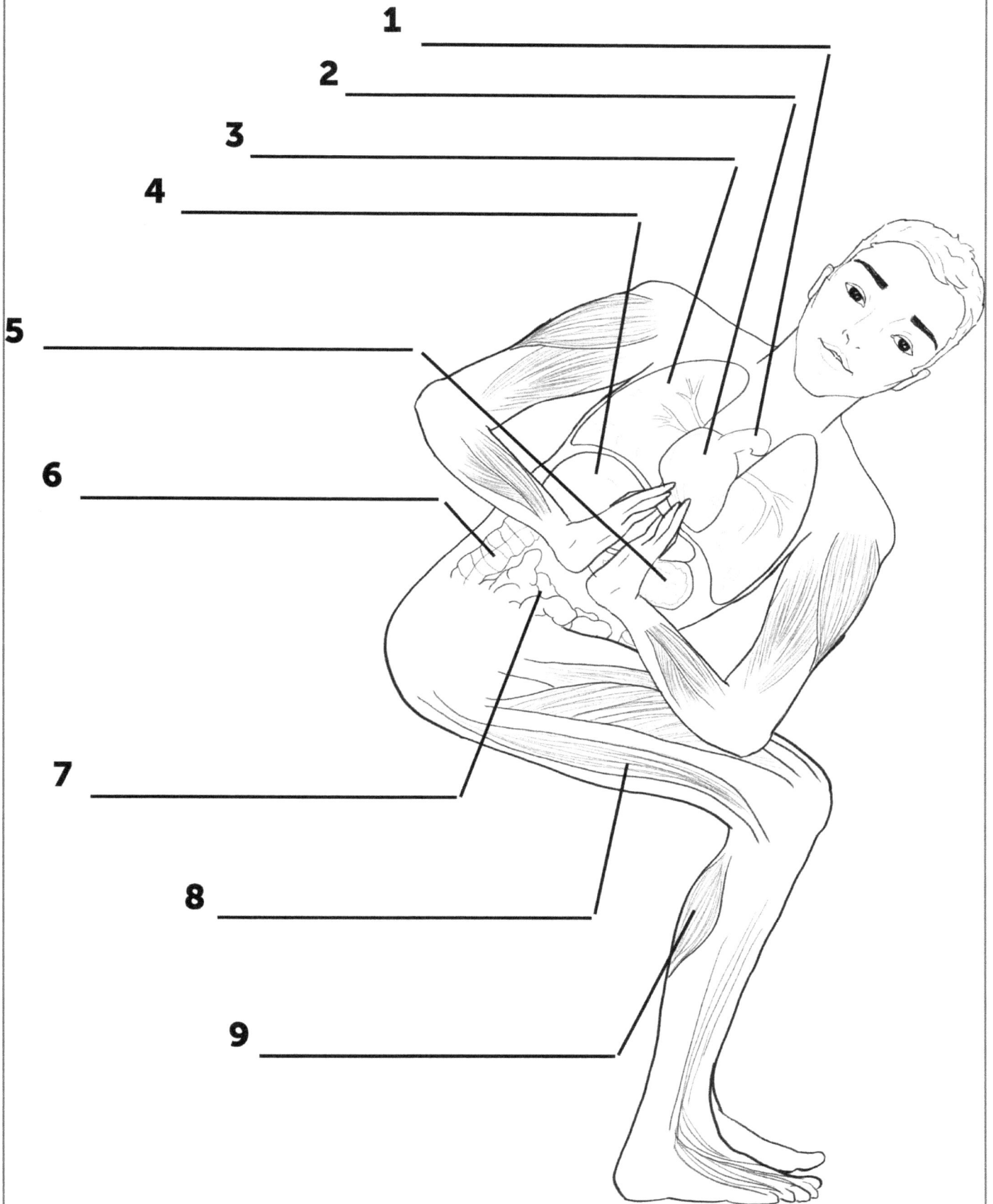

1 _____

2 _____

3 _____

4 _____

5 _____

6 _____

7 _____

8 _____

9 _____

32. TWIST CHAIR POSE

1. AORTA
2. HEART
3. LUNGS
4. LIVER
5. STOMACH
6. ASCENDING COLON
7. COILS OF SMALL INTESTINE
8. HAMSTRINGS
9. GASTROCNEMIUS

33. YOGA RABBIT POSE

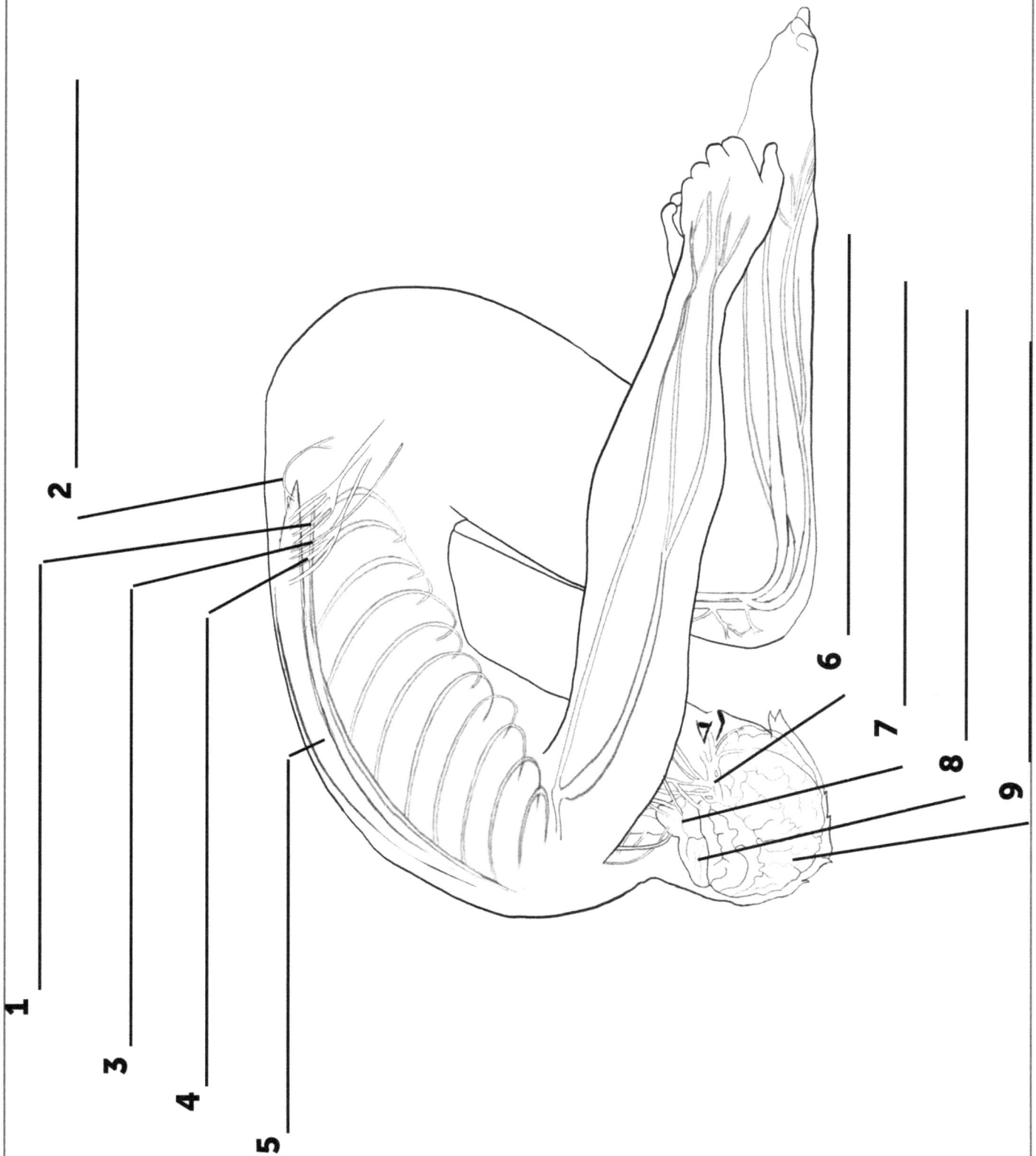

1

2

3

4

5

6

7

8

9

33. YOGA RABBIT POSE

1. SACRAL PLEXUS
2. PUDENDAL NERVE
3. OBTURATOR
4. LUMBAR PLEXUS
5. SPINAL CORD
6. CRANIAL NERVES
7. BRAINSTEM
8. CEREBELLUM
9. CEREBRUM

34. UPWARD PLANK POSE

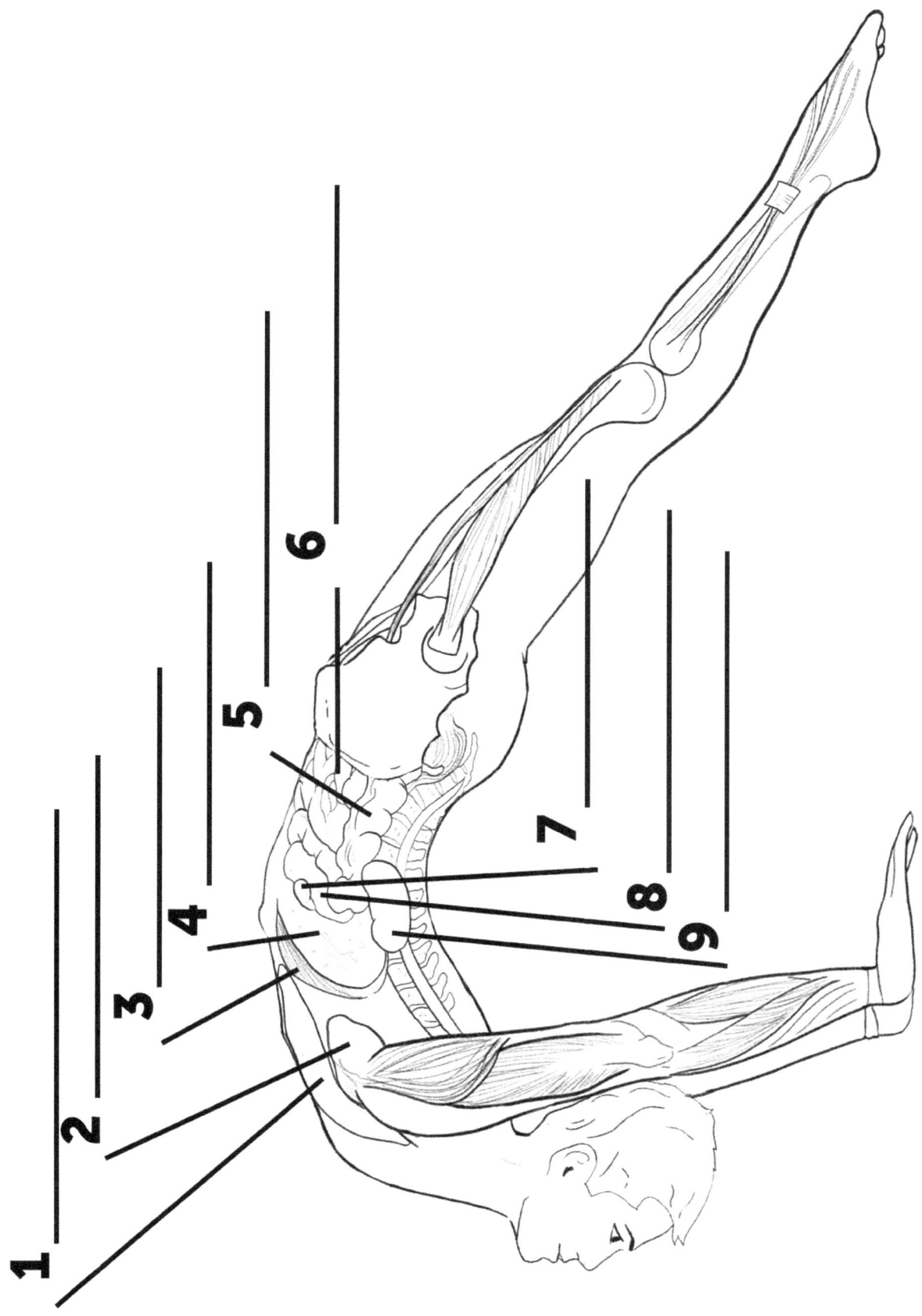

1

2

3

4

5

6

7

8

9

34. UPWARD PLANK POSE

1. LUNGS

2. HEART

3. DIAPHRAGM

4. LIVER

5. ASCENDING COLON

6. COILS OF SMALL INTESTINE

7. GALLBLADDER

8. STOMACH

9. KIDNEY

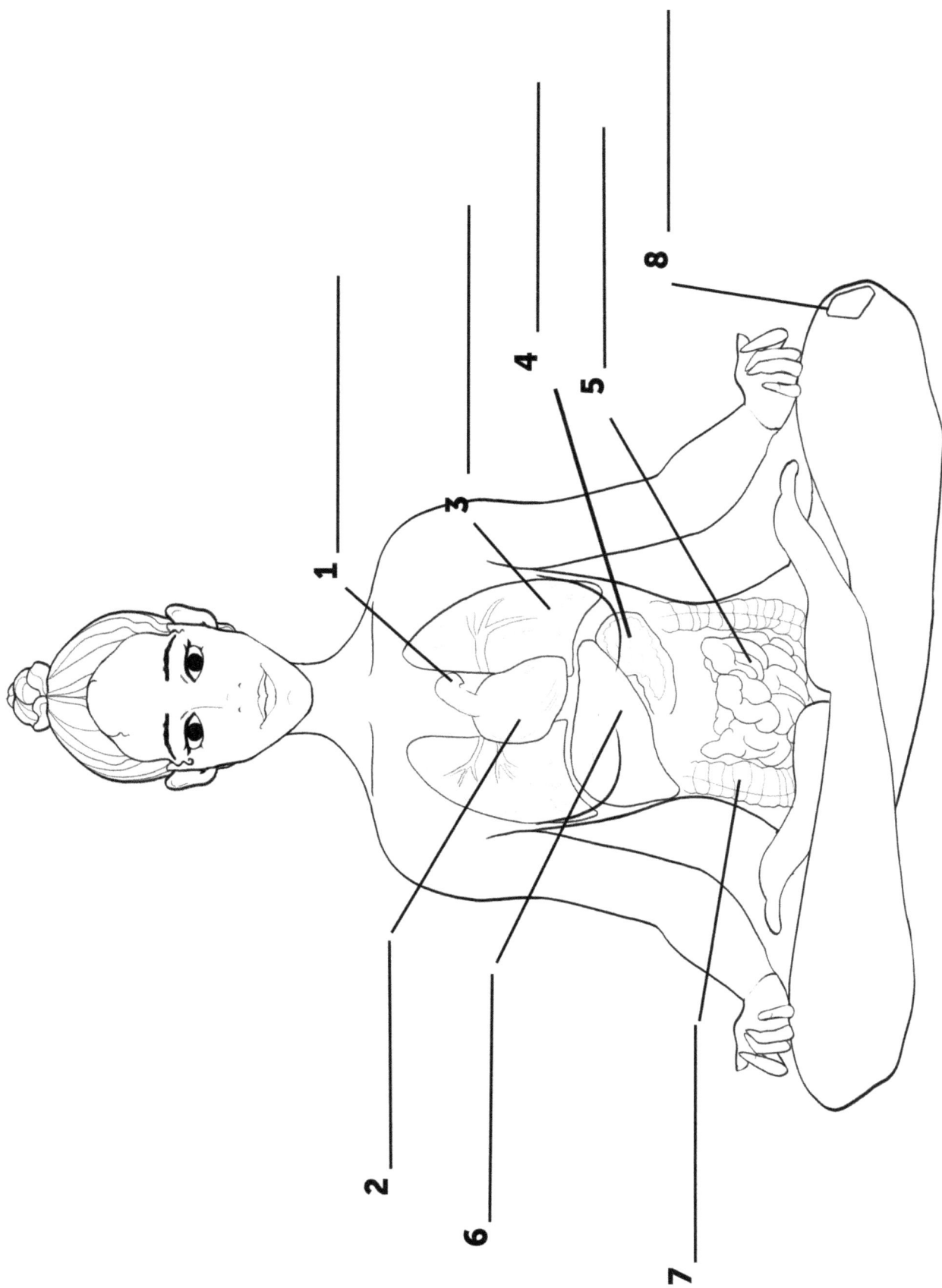

35. LOTUS POSE

1

2

3

4

5

6

7

8

35. LOTUS POSE

1. AORTA

2. HEART

3. LUNGS

4. STOMACH

5. COILS OF SMALL INTESTINE

6. LIVER

7. ASCENDING COLON

8. PATELLA

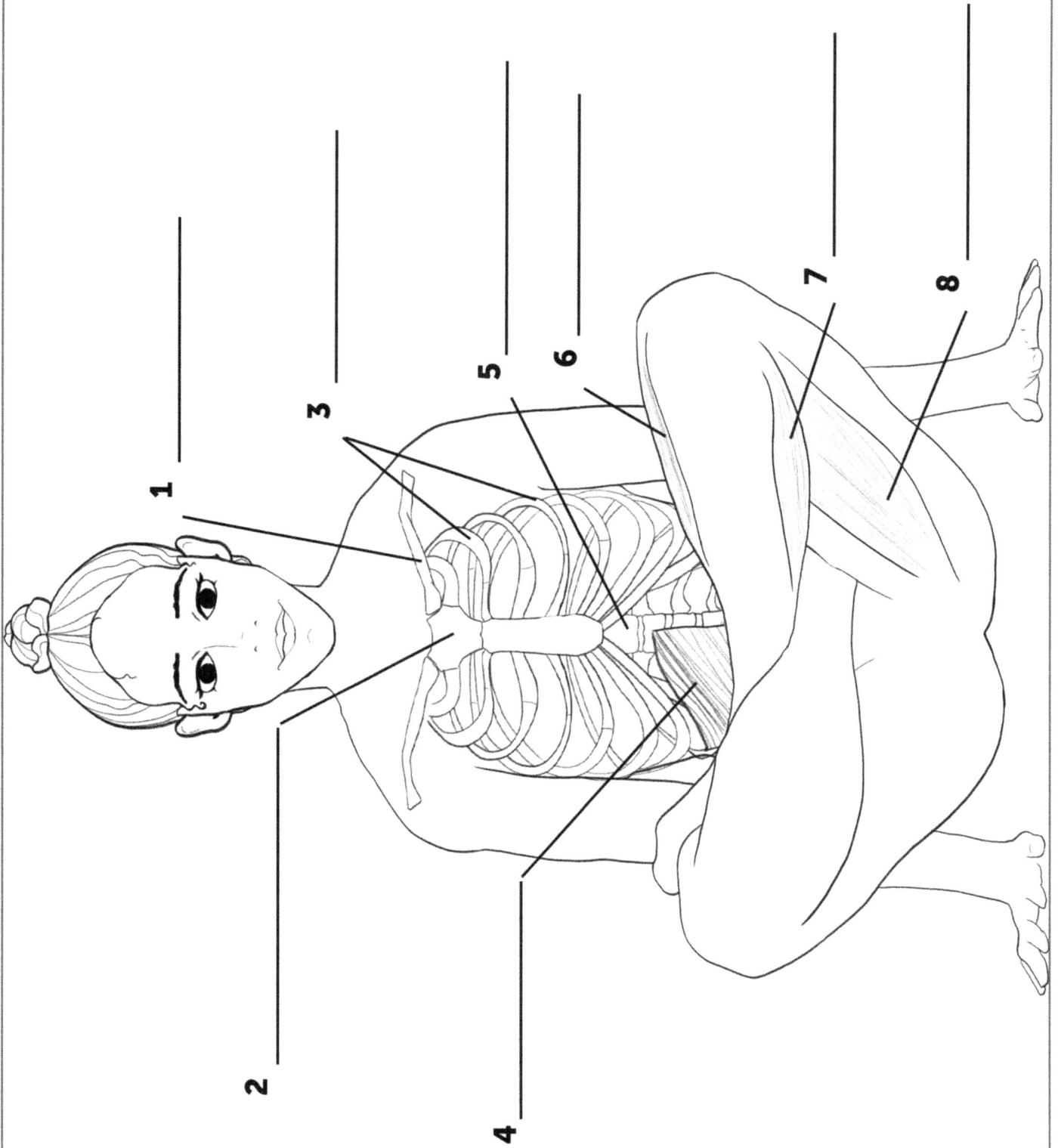

36. SCALE POSE

1

2

3

4

5

6

7

8

36. SCALE POSE

1. COLLARBONE
2. STERNUM
3. RIBS
4. INTERNAL OBLIQUE
5. SPINE
6. GASTROCNEMIUS
7. GASTROCNEMIUS
8. HAMSTRINGS

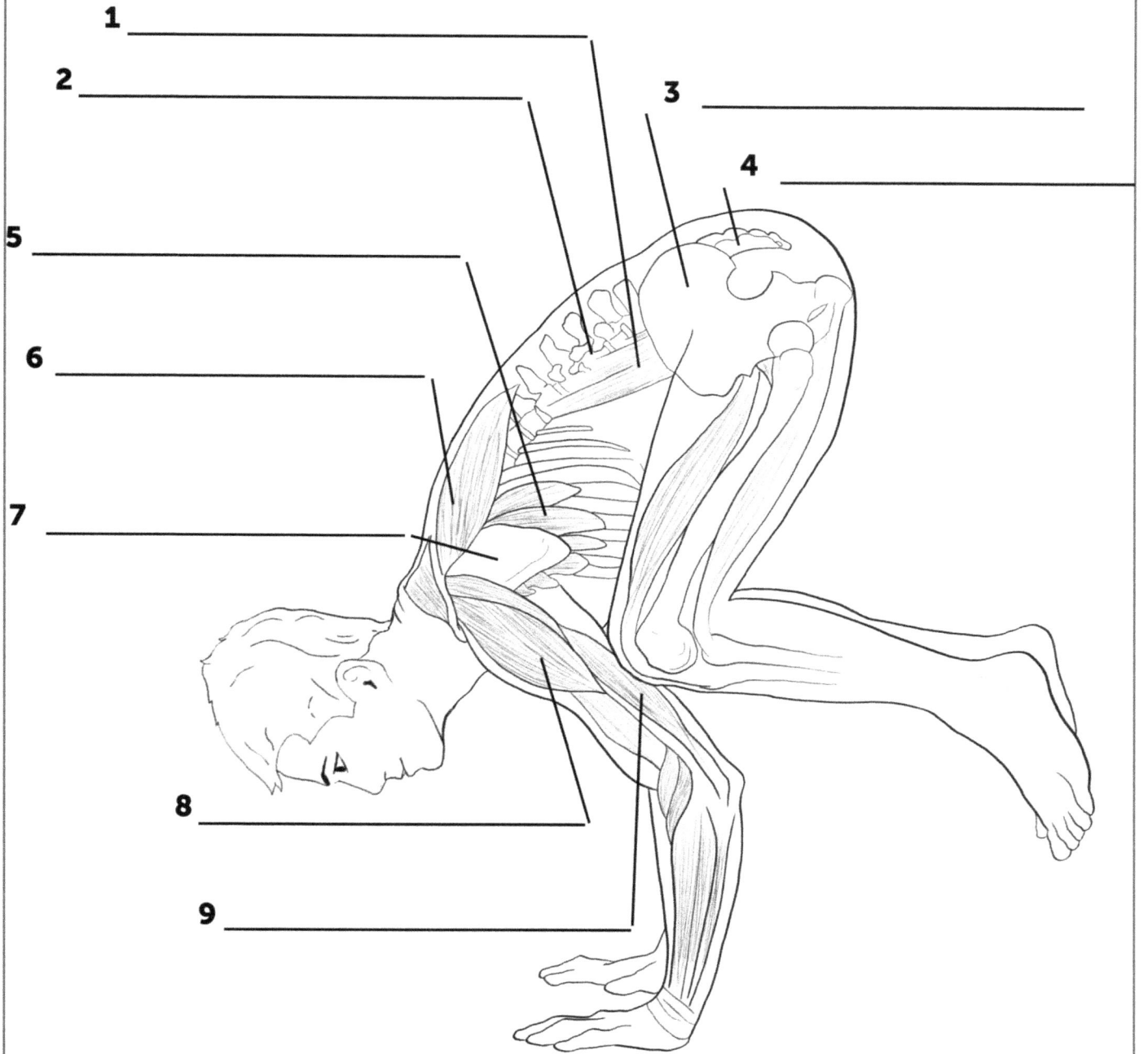

37. CROW POSE

1 _____

2 _____

3 _____

4 _____

5 _____

6 _____

7 _____

8 _____

9 _____

37. CROW POSE

1. PSOAS MAJOR
2. SPINE
3. PELVIS
4. SACRUM
5. SERRATUS ANTERIOR
6. TRAPEZIUS
7. SCAPULA
8. DELTOID
9. TRICEPS BRACHII

38. FOUR LIMBED STAFF POSE

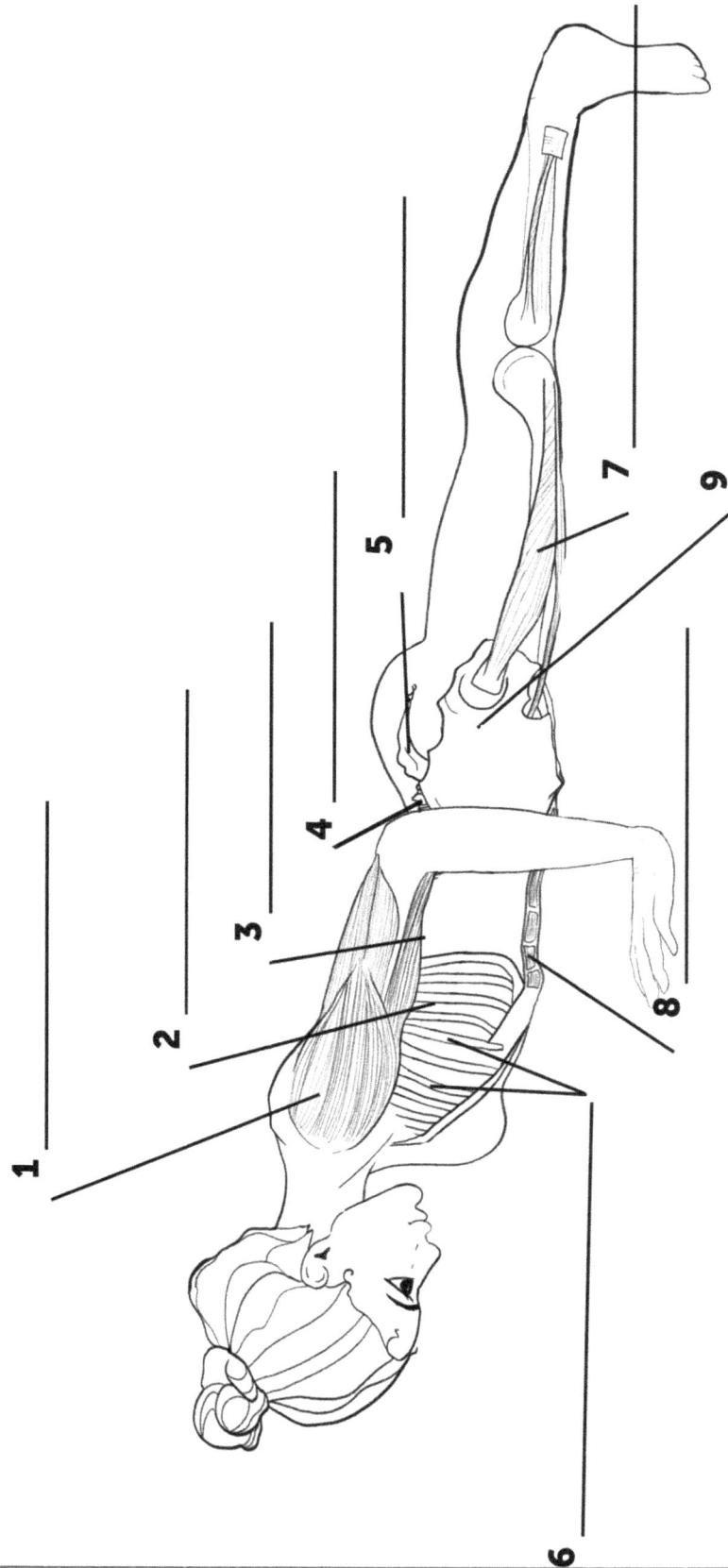

38. FOUR LIMBED STAFF POSE

1. DELTOID
2. RIBS
3. BICEPS BRACHII
4. SPINE
5. SACRUM
6. RIBS
7. RECTUS FEMORIS
8. RECTUS ABDOMINIS
9. PELVIS

39. SIDE CROW POSE

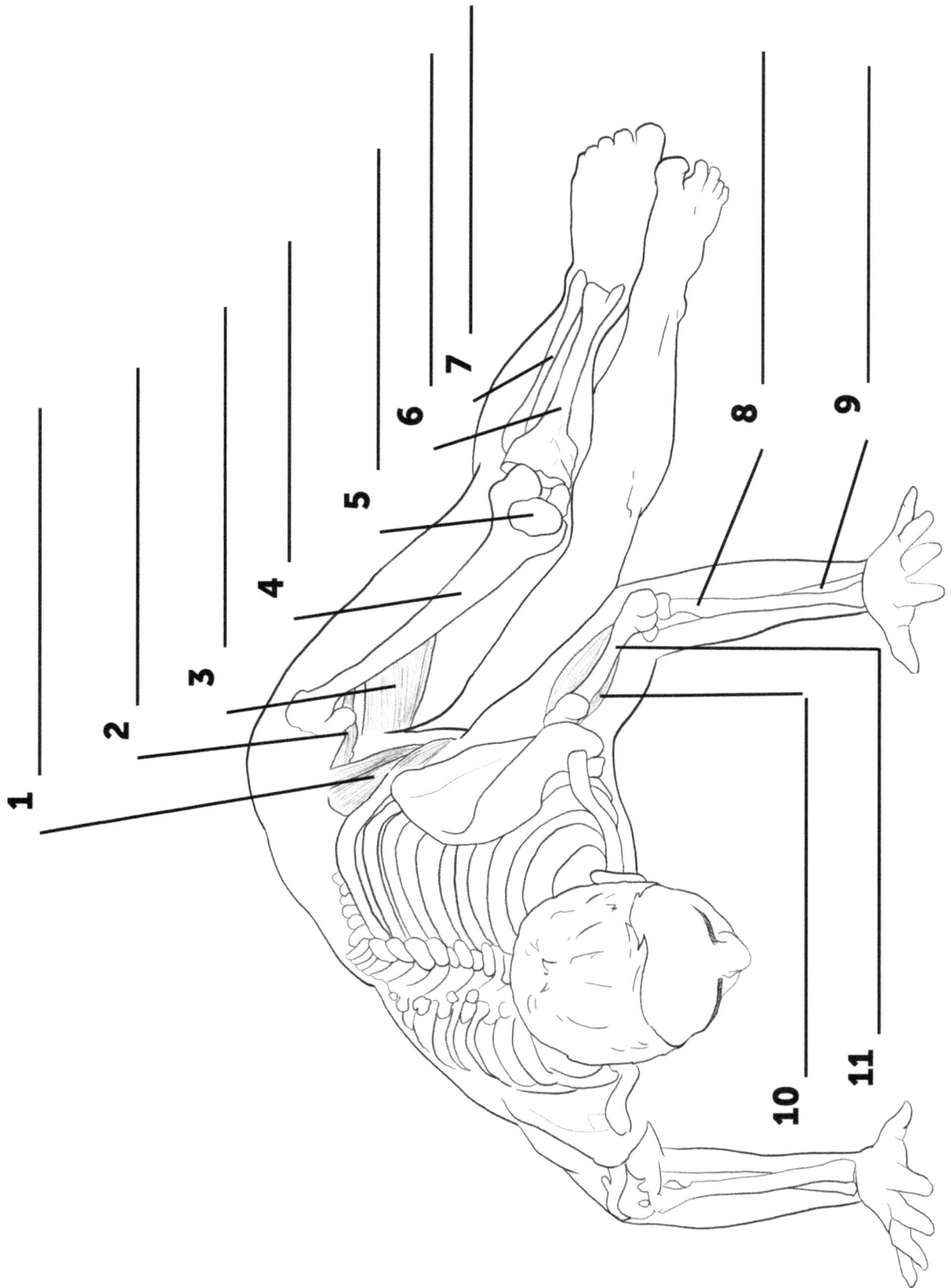

1

2

3

4

5

6

7

8

9

10

11

39. SIDE CROW POSE

1. EXTERNAL OBLIQUE

2. PECTINEUS

3. ADDUCTOR BREVIS

4. FEMUR

5. PATELLA

6. TIBIA

7. FIBULA

8. RADIUS

9. ULNA

10. TRICEPS BRACHII

11. HUMERUS

40. HALF BOAT POSE

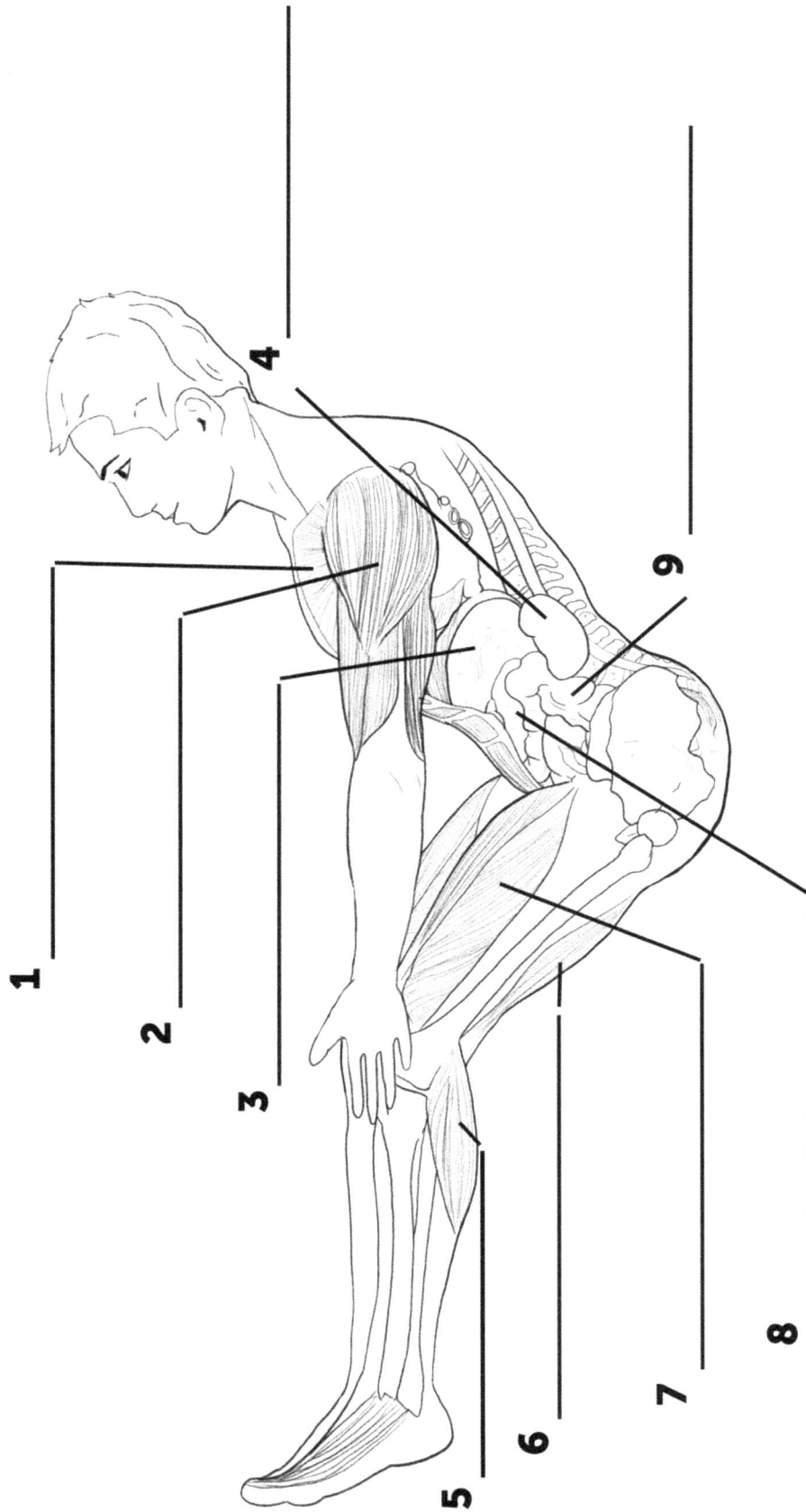

1

2

3

4

5

6

7

8

9

40. HALF BOAT POSE

1. PECTORALIS MAJOR

2. DELTOID

3. LIVER

4. KIDNEY

5. GASTROCNEMIUS

6. HAMSTRINGS

7. QUADRICEPS

8. STOMACH

9. ASCENDING COLON

41. FULL BOAT POSE

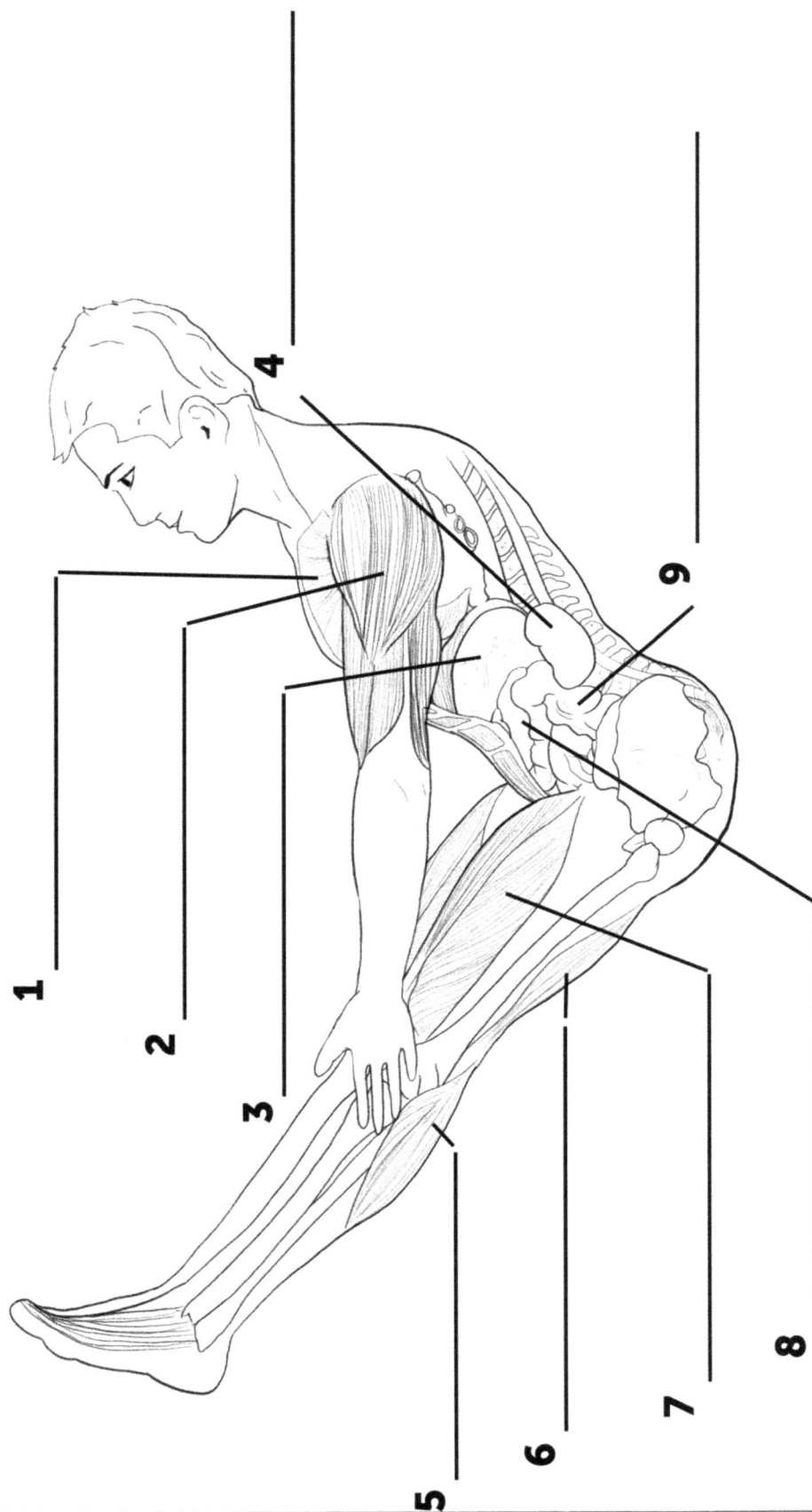

41. FULL BOAT POSE

1. PECTORALIS MAJOR
2. DELTOID
3. LIVER
4. KIDNEY
5. GASTROCNEMIUS
6. HAMSTRINGS
7. QUADRICEPS
8. STOMACH
9. ASCENDING COLON

42. FISH POSE

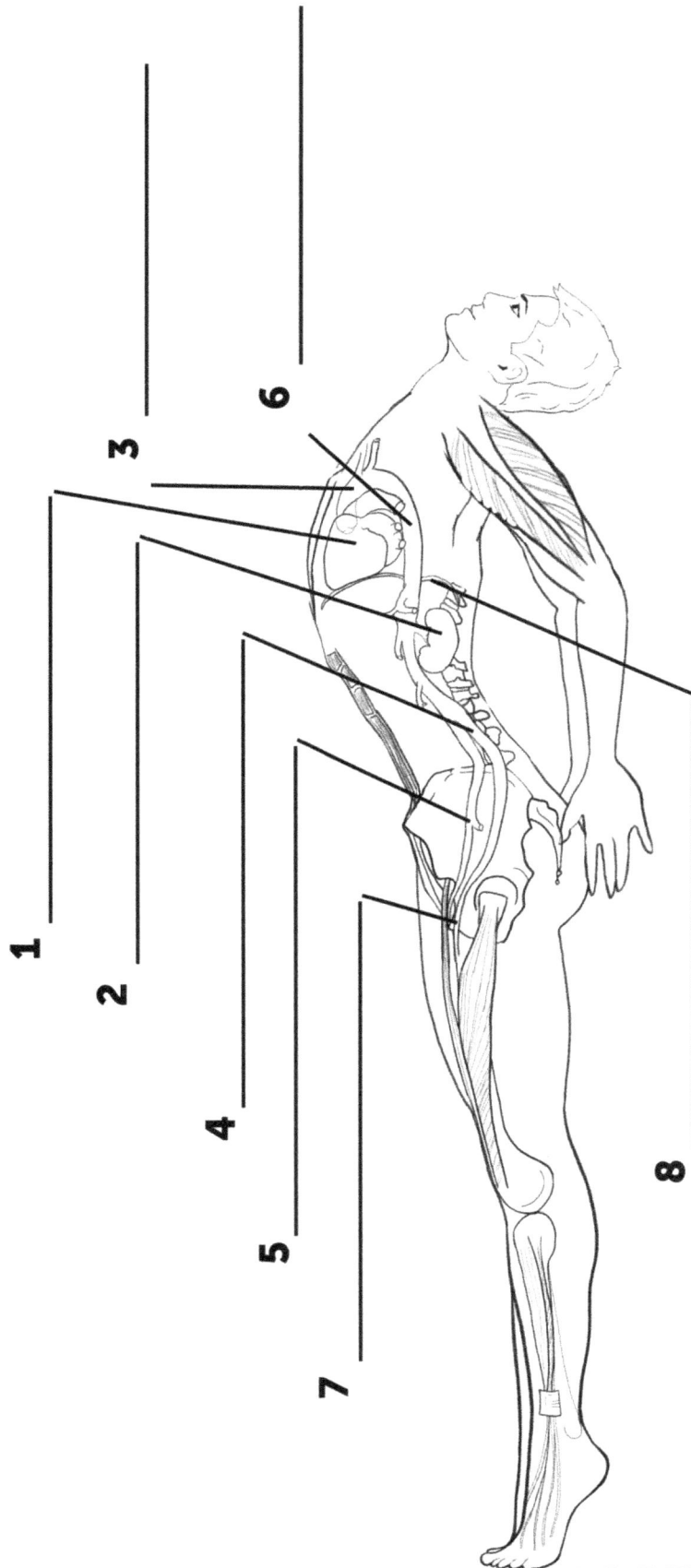

1

2

3

4

5

6

7

8

42. FISH POSE

1. HEART

2. KIDNEY

3. ASCENDING THORACIC AORTA

4. ABDOMINAL AORTA

5. COMMON ILIAC ARTERY

6. DESCENDING THORACIC AORTA

7. FEMORAL ARTERY

8. DIAPHRAGM

43. SUPPORTED HEADSTAND POSE

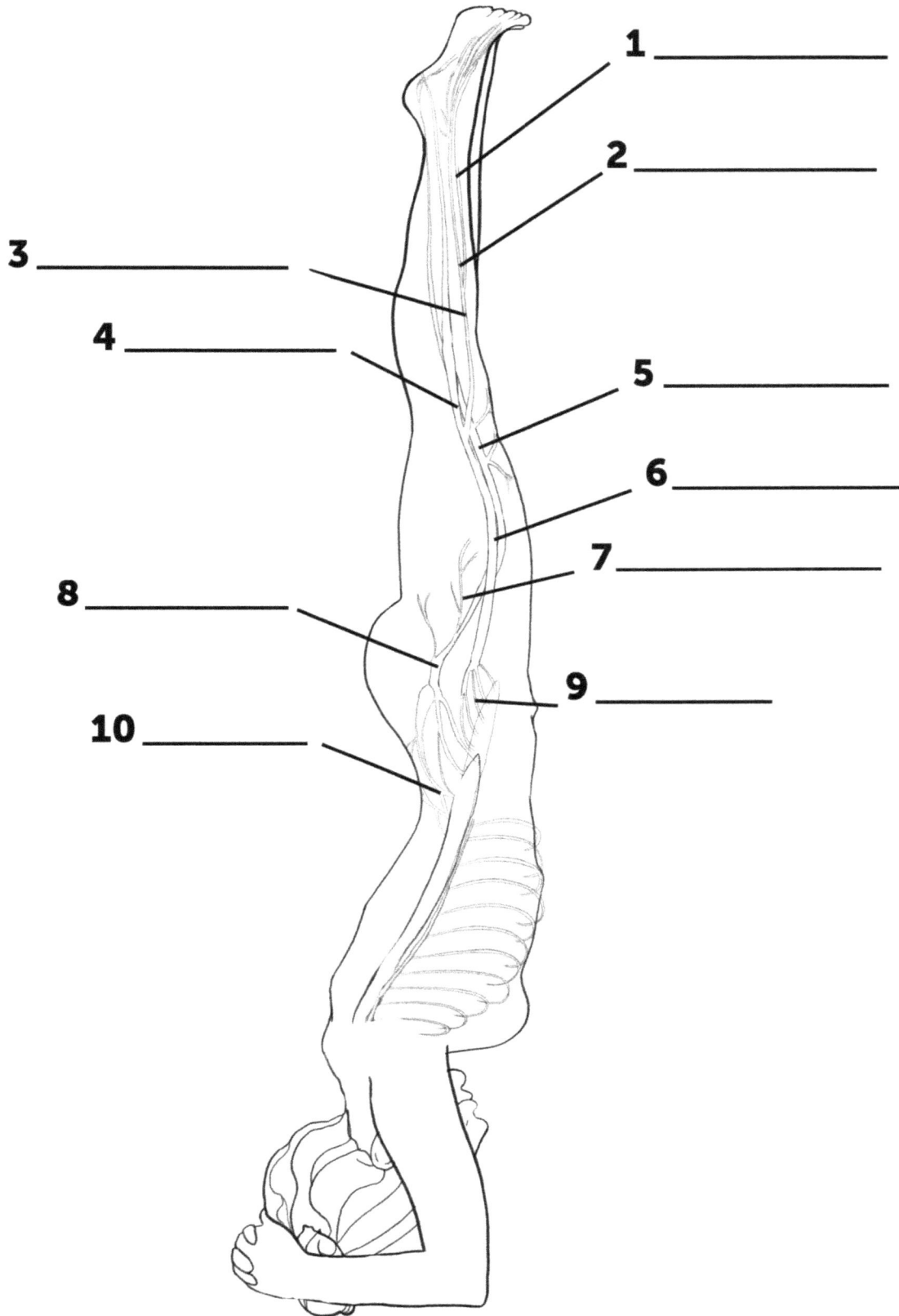

1 _____

2 _____

3 _____

4 _____

5 _____

6 _____

7 _____

8 _____

9 _____

10 _____

43. SUPPORTED HEADSTAND POSE

1. SUPERFICIAL PERONEAL
2. DEEP PERONEAL
3. COMMON PERONEAL
4. TIBIAL
5. SAPHENOUS
6. SCIATIC
7. MUSCULAR BRANCHES OF FEMORAL
8. FEMORAL
9. SACRAL PLEXUS
10. LUMBAR PLEXUS

44. SUPPORTED SHOULDER STAND

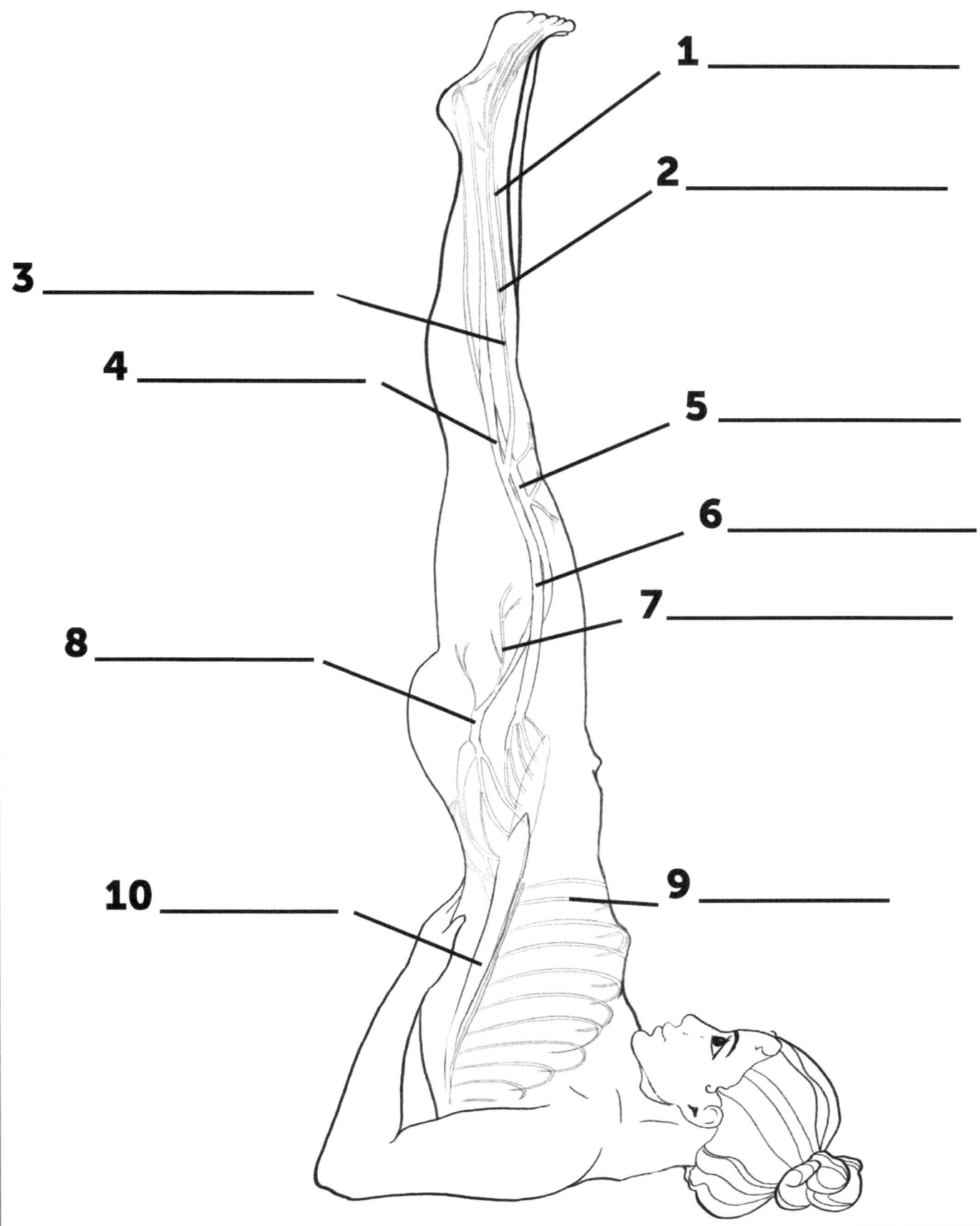

1 _____

2 _____

3 _____

4 _____

5 _____

6 _____

7 _____

8 _____

9 _____

10 _____

44. SUPPORTED SHOULDER STAND

1. SUPERFICIAL PERONEAL

2. DEEP PERONEAL

3. COMMON PERONEAL

4. TIBIAL

5. SAPHENOUS

6. SCIATIC

7. MUSCULAR BRANCHES OF FEMORAL

8. FEMORAL

9. INTERCOSTALS

10. SPINAL CORD

45. PLOW POSE

1

2

3

4

5

6

7

8

9

10

11

12

45. PLOW POSE

1. PELVIS
2. FEMUR
3. HAMSTRINGS
4. GASTROCNEMIUS
5. SOLEUS
6. ERECTOR SPINAE
7. HUMERUS
8. FIBULA
9. TIBIA
10. RADIUS
11. ULNAS
12. TRICEPS BRACHII

46. KNEE-TO-EAR POSE

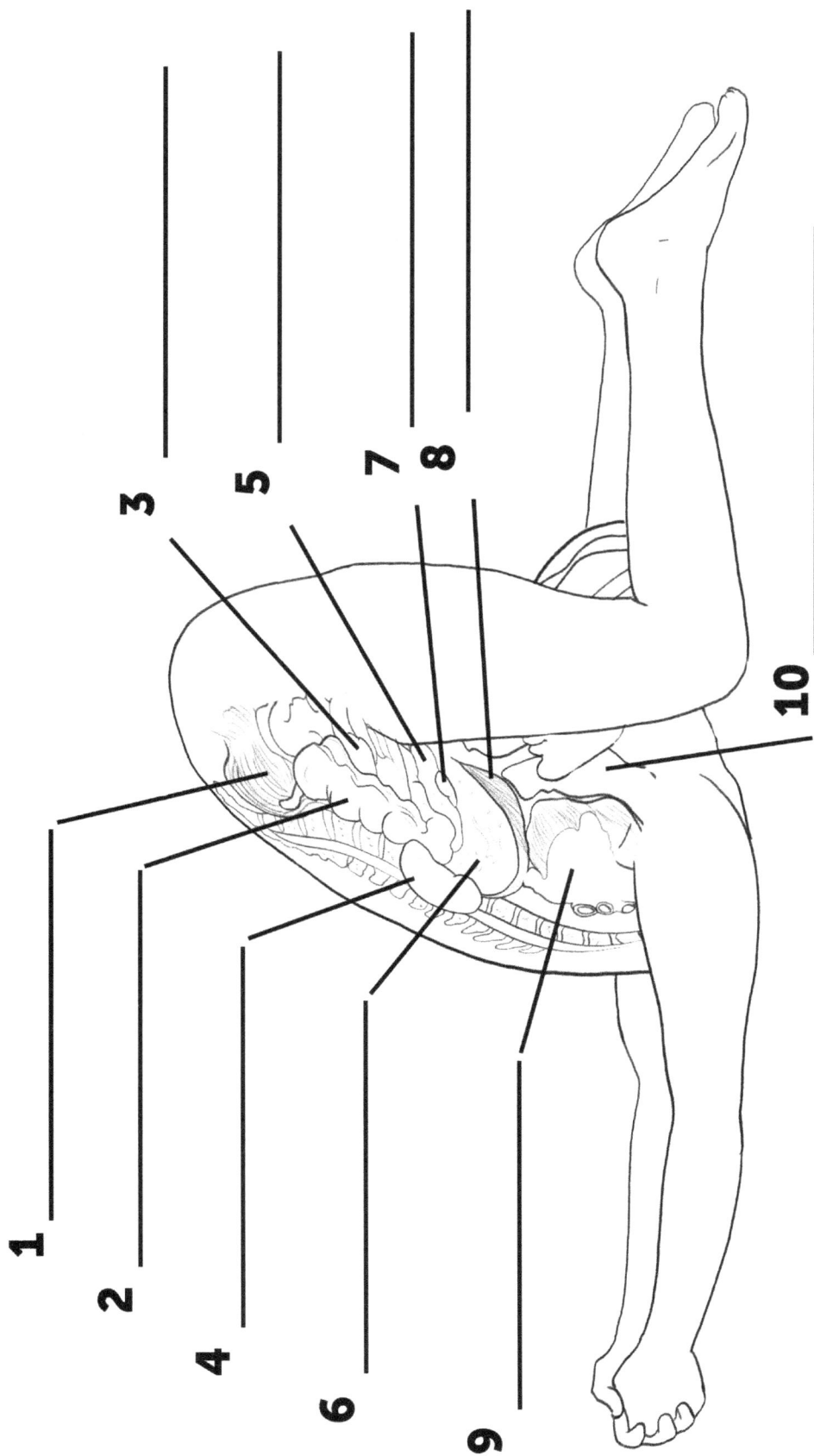

1

2

3

4

5

6

7

8

9

10

46. KNEE-TO-EAR POSE

1.   RECTUM

2.   ASCENDING COLON

3.   COILS OF SMALL INTESTINE

4.   KIDNEY

5.   STOMACH

6.   LIVER

7.   GALLBLADDER

8.   DIAPHRAGM

9.   HEART

10. LUNGS

47. HALF-MOON POSE

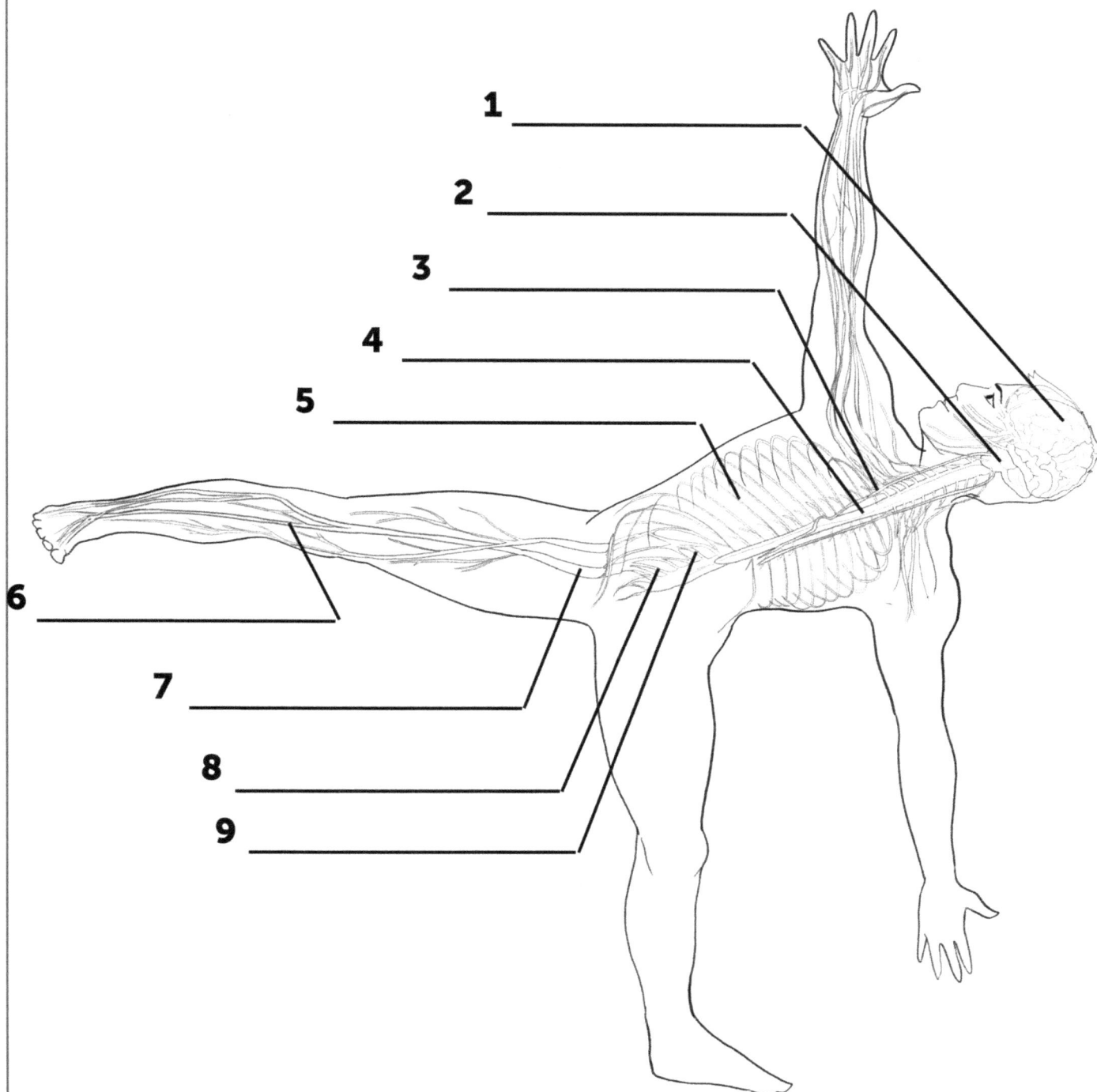

1 _____

2 _____

3 _____

4 _____

5 _____

6 _____

7 _____

8 _____

9 _____

47. HALF-MOON POSE

1. CEREBRUM
2. BRAINSTEM
3. BRACHIAL PLEXUS
4. SPINAL CORD
5. INTERCOSTALS
6. TIBIAL
7. SCIATIC
8. SACRAL PLEXUS
9. LUMBAR PLEXUS

48. COMPASS POSE

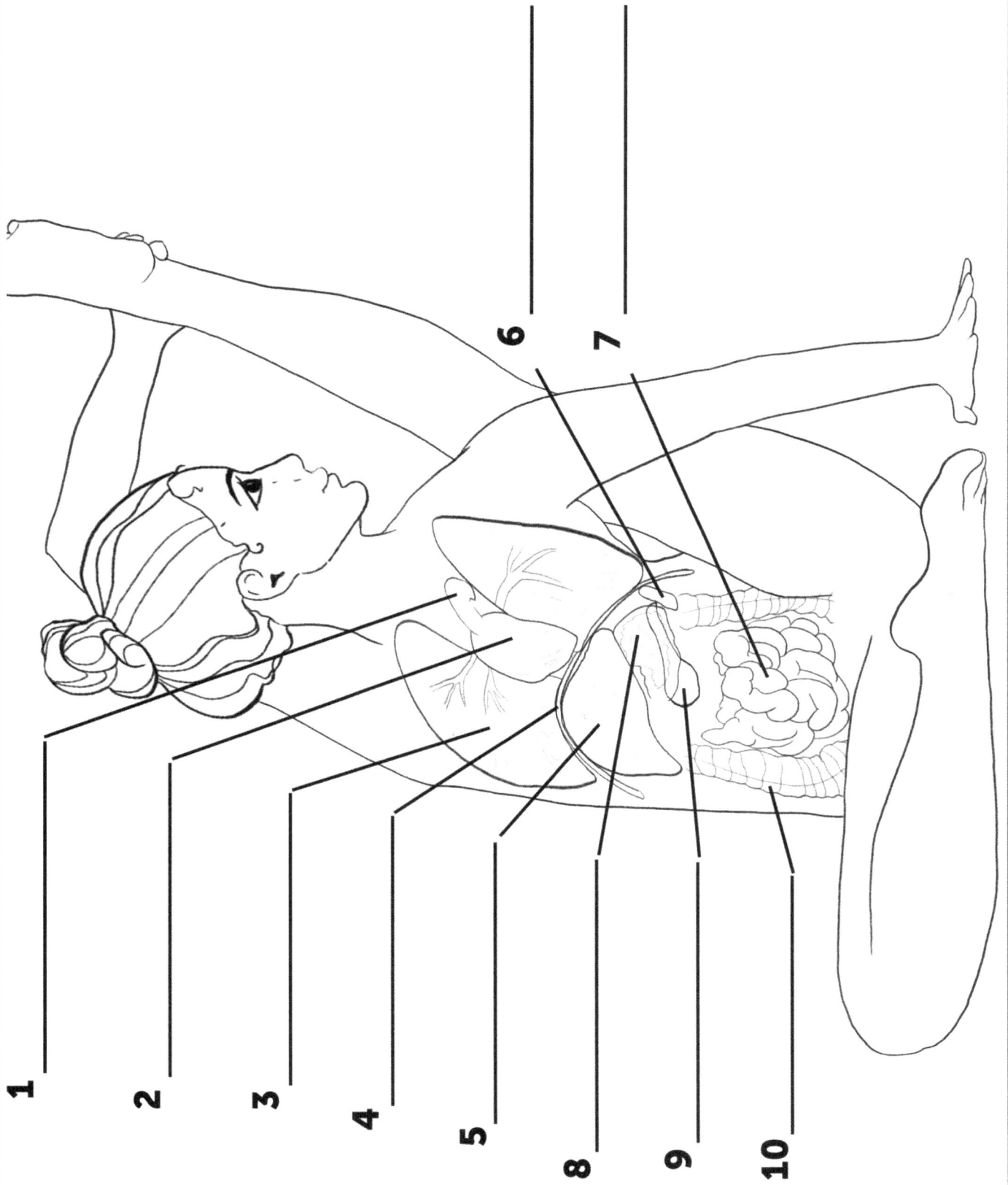

1

2

3

4

5

6

7

8

9

10

48. COMPASS POSE

1. AORTA
2. HEART
3. LUNGS
4. DIAPHRAGM
5. LIVER
6. SPLEEN
7. COILS OF SMALL INTESTINE
8. STOMACH
9. PANCREAS
10. ASCENDING COLON

49. TWISTED HEAD-TO-KNEE POSE

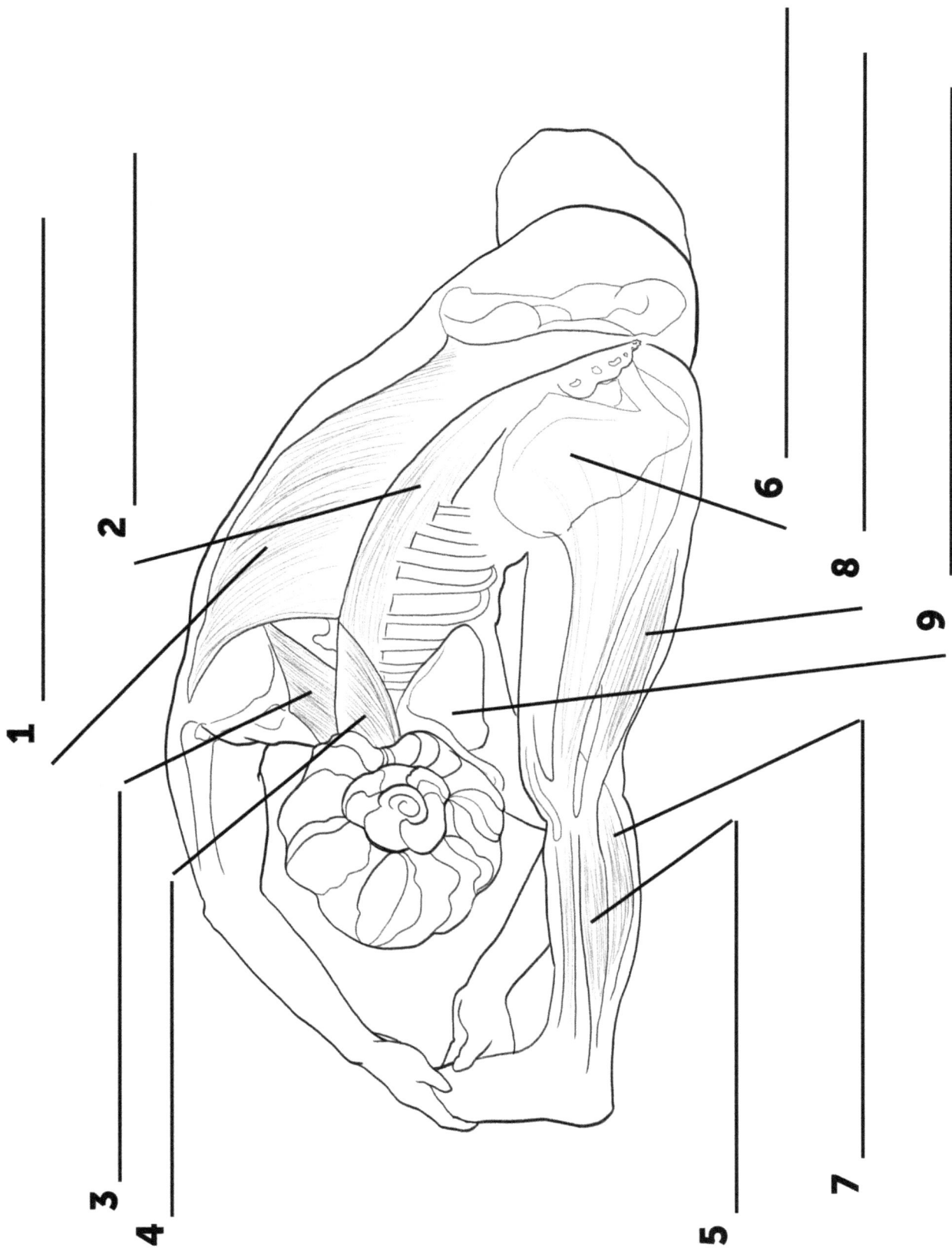

1

2

3

4

5

6

7

8

9

49. TWISTED HEAD-TO-KNEE POSE

1. LATISSIMUS DORSI

2. ERECTOR SPINAE

3. RHOMBOIDS

4. TRAPEZIUS

5. SOLEUS

6. PELVIS

7. GASTROCNEMIUS

8. HAMSTRINGS

9. SCAPULA

50. STANDING SPLIT POSE

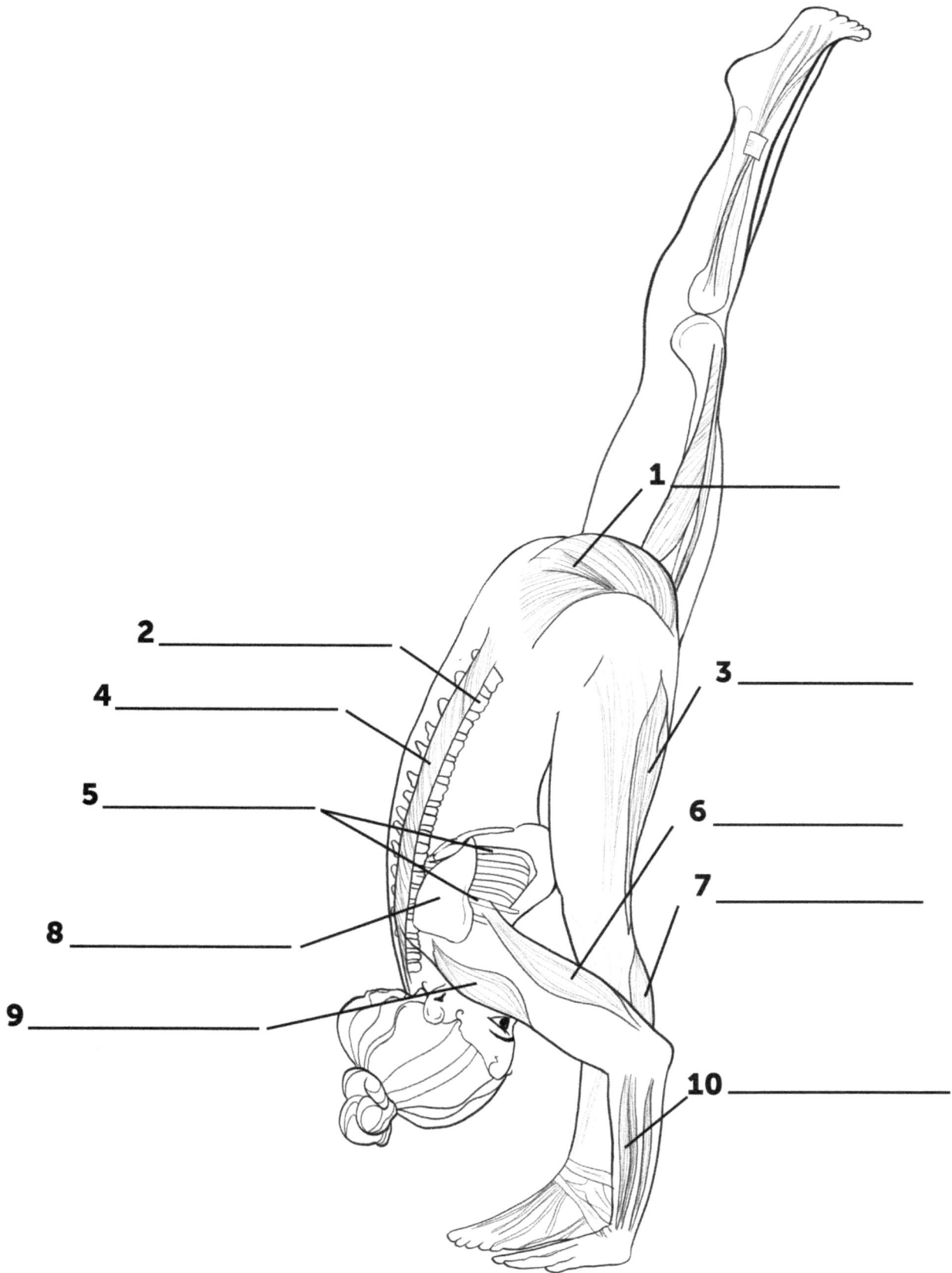

1 _____

2 _____

4 _____

3 _____

5 _____

6 _____

7 _____

8 _____

9 _____

10 _____

50. STANDING SPLIT POSE

1. PIRIFORMIS

2. SPINE

3. HAMSTRINGS

4. ERECTOR SPINAE

5. RIBS

6. TRICEPS BRACHII

7. GASTROCNEMIUS

8. SCAPULA

9. DELTOID

10. PRONATORS

51. ARCHER POSE

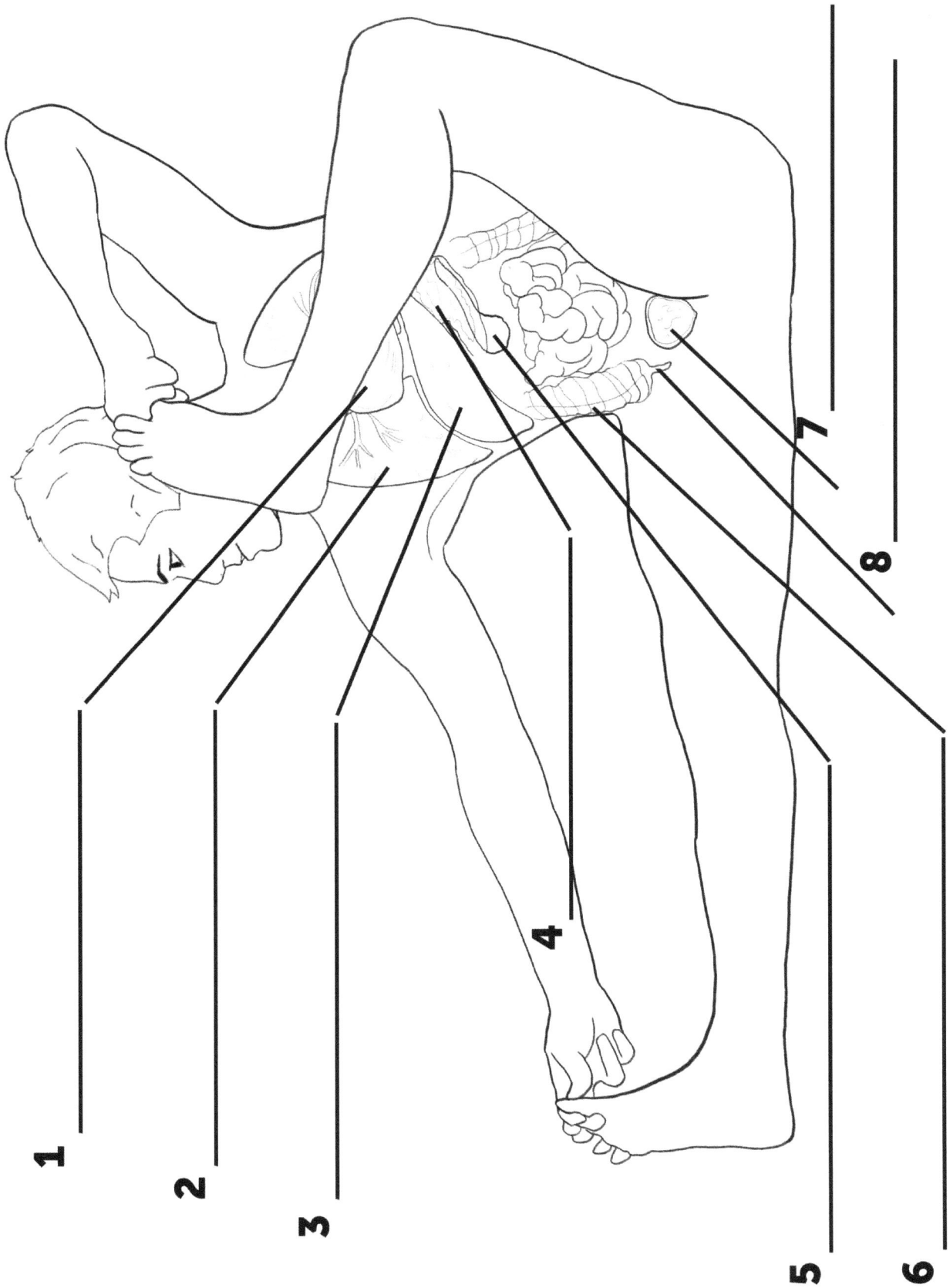

1

2

3

4

5

6

7

8

51. ARCHER POSE

1. HEART
2. LUNGS
3. LIVER
4. STOMACH
5. PANCREAS
6. ASCENDING COLON
7. URINARY BLADDER
8. APPENDIX

52. YOGA HANDSTAND POSE

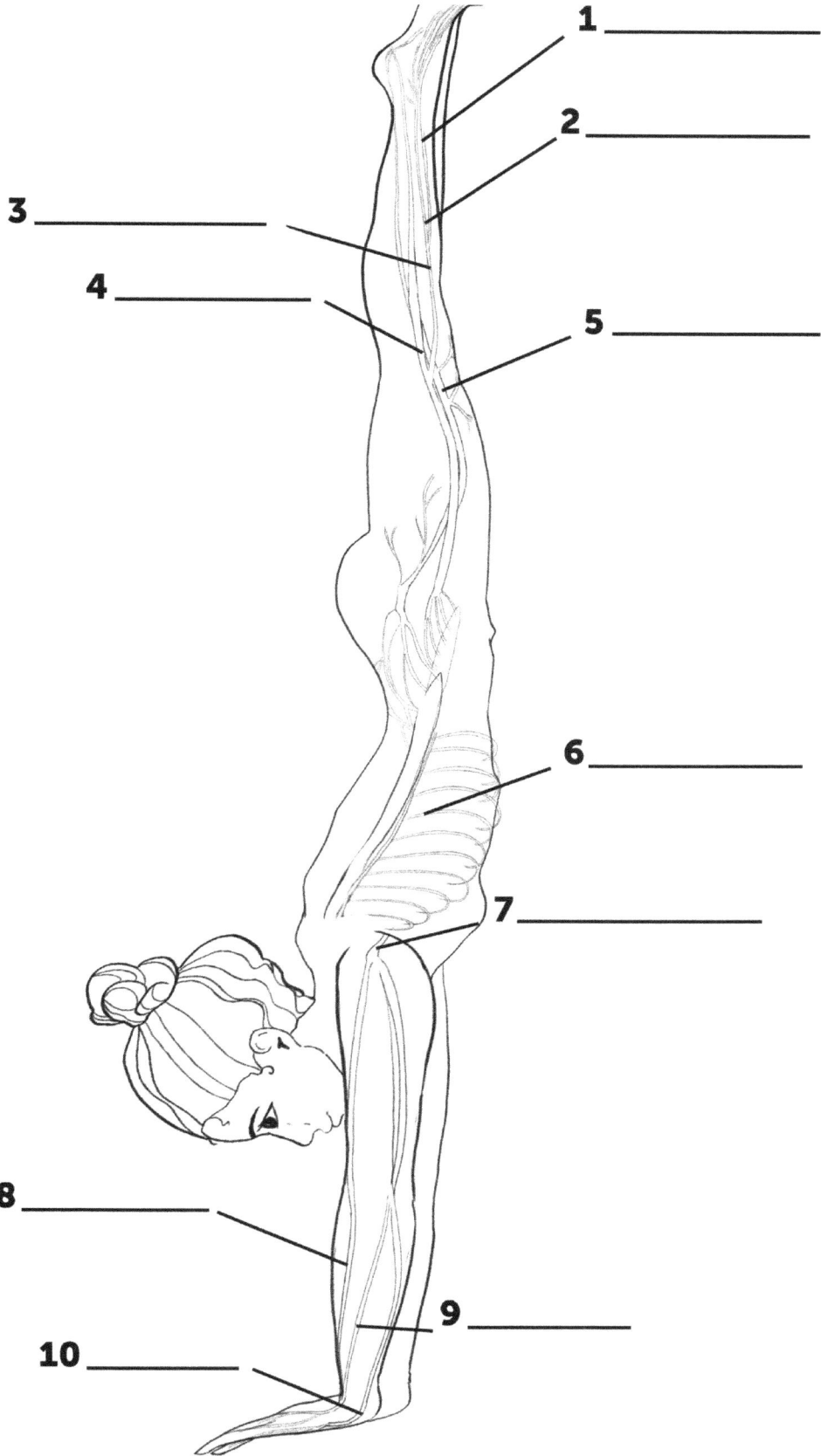

1 _____

2 _____

3 _____

4 _____

5 _____

6 _____

7 _____

8 _____

9 _____

10 _____

52. YOGA HANDSTAND POSE

1. SUPERFICIAL PERONEAL

2. DEEP PERONEAL

3. COMMON PERONEAL

4. TIBIAL

5. SAPHENOUS

6. INTERCOSTALS

7. BRACHIAL PLEXUS

8. RADIAL

9. MEDIAN

10. ULNAR

53. ELEPHANT TRUNK POSE

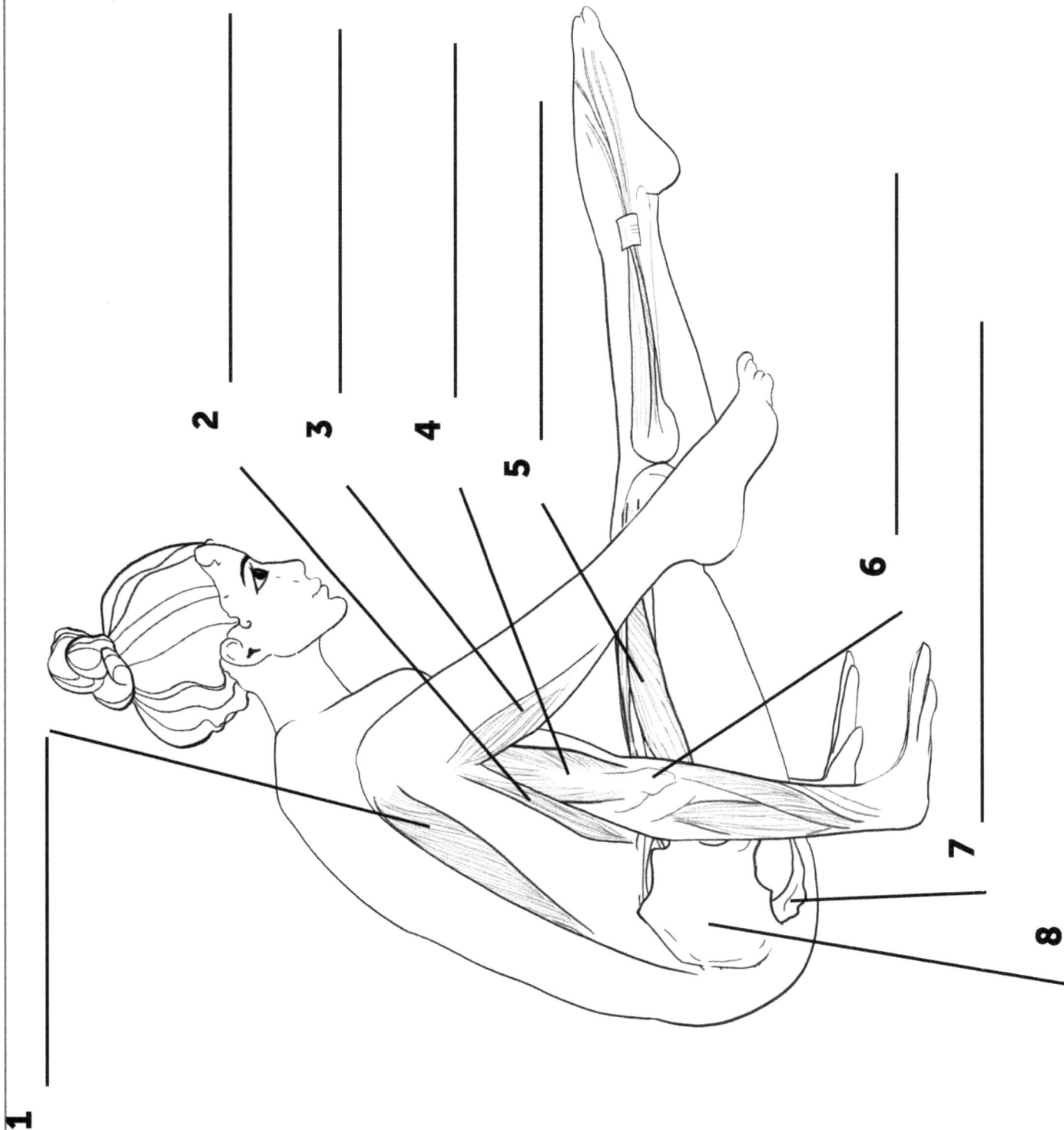

1

2

3

4

5

6

7

8

53. ELEPHANT TRUNK POSE

1. RECTUS FEMORIS
2. HAMSTRINGS
3. GASTROCNEMIUS
4. TRICEPS BRACHII
5. QUADRICEPS
6. ELBOW
7. SACRUM
8. PELVIS

www.ingramcontent.com/pod-product-compliance
Lightning Source LLC
Chambersburg PA
CBHW051342200326
41521CB00015B/2588